LESLEY J. ROGERS holds a Personal Chair at the University of New England, Australia. In the two decades since completing a doctor of philosophy in ethology at the University of Sussex, she has published widely in leading scientific journals and has been author or co-author of three other books on animal behaviour and development. Lesley's research bridges the disciplines of neuroscience and behaviour, and she works in both the field and the laboratory. In 1987, the University of Sussex awarded her a Doctor of Science.

MINDS OF THEIR OWN

Thinking and awareness in animals

Lesley J. Rogers

ALLEN & UNWIN

Copyright © Lesley J. Rogers 1997

First published in 1997 by
Allen & Unwin
9 Atchison Street
St Leonards NSW 2065
Australia
Phone: (61 2) 9901 4088
Fax: (61 2) 9906 2218
E-mail: frontdesk@allen-unwin.com.au
URL: http://www.allen-unwin.com.au

National Library of Australia
Cataloguing-in-Publication entry:

Rogers, Lesley J. (Lesley Joy), 1943– .
    Minds of their own: thinking and awareness in animals.

Includes index.
ISBN 1 86448 504 3.

1. Animal psychology. 2. Animal behavior. 3. Consciousness in animals. I. Title.

156.3

Set in 10/12pt Plantin by DOCUPRO, Sydney
Printed by South Wind Productions, Singapore

10 9 8 7 6 5 4 3 2 1

CONTENTS

LIST OF FIGURES

ACKNOWLEDGEMENTS

I am most indebted to Professor Gisela Kaplan for not only correcting a draft of this manuscript but also making challenging comments on it. I also thank her for proof reading and, especially, for teaching me much more about birds in all of their varieties than I had already learnt in nearly thirty years of studying chickens. Her rescue and care of a wide variety of injured Australian birds brought me into close contact with species that I would never otherwise have encountered at such close quarters. Ten years ago, she also took me from my laboratory into the rainforests to study orang-utans, a research project that we have shared since then. Although I have always been interested in the complex behaviour of animals, it was the contact with orang-utans that turned my attention towards the topic of this book.

I am also grateful to numerous discussion group meetings with my postgraduate students. Many of the ideas that I have developed in the book arose from discussions with my colleagues and friends Professors Richard Andrew, Mike Cullen, Jeannette Ward, John Bradshaw, Allen Gardner and the late Beatrix Gardner, and also Doctors Giorgio Vallortigara, Christopher Evans and Nicola Clayton. Some of these may agree with some of the things that I have said, others will not. I am also most grateful for the positive encouragement and other assistance of my publisher Ian Bowring. Thank you also to Andrew Robins for assistance in collecting reference material.

I would like to dedicate this book to Jenny, Luke and Julie, who happen to be Rhodesian ridgebacks. They sat at my feet through the entire writing phase of this book. I realise that in doing so I run the risk of being labelled a sentimentalist rather than a scientist, but I do not think that these labels are in conflict.

CHAPTER ONE

NOT SIMPLY MACHINES

Do animals have ideas and do they think about objects that they cannot see or about situations that have occurred in the past? Do they consciously make plans for the future or do they simply react unthinkingly to objects as they appear and situations as they arise? Are animals aware of themselves and of others or is this an ability unique to humans? All of these questions have bearing on whether animals have consciousness or not.

We live at a time when the debate about consciousness in animals has taken a new turn and may have greater meaning than ever before. A number of seemingly separate lines of thinking have come together to lead us to consider the issue afresh. Some computers are said to have 'intelligence', and they can 'learn' in ways that we never thought possible a decade ago. There is every possibility that machines of the future will process information in an even more human-like way. It is, of course, debatable whether they will be able to 'think' like humans and, as far as I know, only very few people expect them to become conscious. At the same time as these sophisticated computers have been developed, we have realised that, although humans have consciousness, at least some of our behaviour is carried out quite unconsciously. We sometimes perform apparently rather complex learnt sequences of behaviour without being fully aware of what we are doing, rather like a sleep walker. Of course, this unconscious, or more often half conscious, control of behaviour usually occurs for only very short periods of time, but it can startle us when we 'wake up to it'. All of us must have, on occasions, found ourselves driving on a familiar route, making turns and avoiding traffic, without being fully conscious of the decisions that we are making along the way.

Unconscious thoughts and memories may also influence our conscious behaviour. We have known this from the time of Freud. To use the terminology of Freudian psychology, our conscious behaviour may be influenced by *sub*conscious memories and drives of which we have no awareness at the time. Of course, the existence of these underlying thoughts remains a matter of surmise because they are concealed by their very subconsciousness.

Animals too may perform some behavioural acts unconsciously. Sometimes my dog attempts to bury a biscuit in her bed by wiping the mattress with her snout in a repeated and typical movement that would have buried the biscuit were it on soft soil. I have noticed her also wiping the ground in the same way after she has regurgitated food even though it is on a hard surface. She rakes 'imaginary' soil and is, seemingly, unconscious of her lack of achievement in covering the material. Some readers may see this behaviour as instinctual, meaning inherited or preprogrammed in the genes that are passed on from one generation to the next. However, even learnt patterns of behaviour can be performed in such seemingly mechanical ways. We refer to them as habits.

How much of animal behaviour is automatic? When and how does information processing in the brain become conscious? Consciousness is one of the characteristics that we have attributed to ourselves alone amongst animals. There are also other characteristics that we have used to separate ourselves from other creatures. These include language, use of symbols in art, and tool use. We have also seen our superiority in terms of walking in an upright posture (bipedalism), having a lateralised brain and being more intelligent.

2

Considerable numbers of people in the Western world believe that animals are little more than machines, albeit more or less complex ones depending on the species. However, there is increasing debate about awareness in animals and much new information relevant to this debate has come to light. Following on from this, there is a new interest in the welfare of animals and even discussion of the rights of animals. The outcomes of the present debate will determine how we treat animals in the laboratory, in agriculture, in zoos and in our homes. Far from being an esoteric debate, it is central to the current concerns about animal welfare and animal rights. For example, do animals experience pain and suffering as we humans do? We refer to the ability to feel pain as being sentient. Do animals feel love, hatred, happiness, sadness and so on as we do? All of these feelings, in one way or another, reflect a degree of consciousness or awareness.

What do we mean by consciousness? To most people, to be conscious means to be aware of oneself as well as to be aware of others, but there is no agreed, single definition of consciousness. As mentioned already, to be able to think about things not present in the immediate environment is also considered to be an aspect of consciousness and so is the ability to feel and express human-like emotions. Subjectively, we have no great difficulty in knowing for ourselves what consciousness is, but it is not so simple to know about the consciousness of another human, let alone another animal.

The lack of a single definition for consciousness is one of the reasons that many scientists say they do not want to study it. If you cannot define what you are looking for there is no way of studying it objectively. Consciousness is so subjective that scientists might speculate in their spare moments or in conversations with each other whether it exists but very few of them have conducted experiments or made observations that attempt to measure this mysterious thing we call 'consciousness'. At the same time as philosophers debate whether we all experience 'red roses'

3

in the same way and whether we can ever really know if persons other than ourselves are not zombies, scientists who study the structure and function of nerve cells in the brain (i.e. neuroscientists) are prepared to accept that all humans are conscious and to proceed to speculate about where in the brain they might find the elusive neural circuits in which consciousness resides.

Recently, some neuroscientists have started to look for electrical events that may underlie consciousness by recording from the brains of animals. These particular scientists believe that consciousness, of some kind, exists in animals, otherwise they would be unable to conduct their experiments. They believe that we will, one day, explain consciousness by the standard methods of neuroscience and psychology, even though it is out of reach at present. Many of these scientists are reductionists, as they reduce explanations of consciousness to molecular and electrical events.

Others say that we will never be able to explain consciousness by these lower level events. People of this opinion say that, although consciousness may emerge from physical processes of the brain, the firing of nerve cells or similar events, it is something intangible that will never be reached even by new tools or new discoveries. In this case, consciousness would be an epiphenomenon beyond observation and measurement. Personally, I doubt whether this is correct but nor do I think that consciousness can be explained *only* in terms of physical and molecular processes.

Even if it is impossible to measure consciousness as some sort of physical entity (e.g. as oscillations in the cerebral cortex of the brain) now or in the future, we may be able to assess its presence or absence by observation of the behaviour of individuals. As Marian Dawkins of Oxford University has said, if consciousness has a function, it should affect the behaviour of individuals that have it. That is, by observing their behaviour we should be able to detect signs indicating whether they are conscious, even though we might not be able to measure consciousness itself. This approach provides a starting point for us. Consciousness might be manifested in a range of behaviours and we might be able to find patterns of behaviour that indicate consciousness. This way a single definition of consciousness is not needed before we start the search for signs of consciousness. As Donald Griffin has said, it is a mistake to use the absence of a definition as a reason for not investigating whether animals can think and might be conscious.

There are new aspects of the debate about consciousness, but the issue of consciousness in animals has had a very long history. The Greek philosopher Aristotle proposed that humans possess the power to reason, whereas animals do not. Accordingly, nonhuman animals simply act on the basis of innate knowledge, following a set of inherited rules or programs for behaviour without thinking and with little ability, if any, to adapt to new situations. In the seventeenth century, René Descartes described humans as conscious beings and animals as automata, machines. There were many others of his time who thought likewise. Descartes was fascinated by the functioning of the human body and made great advances in the sciences of anatomy and physiology. He was also interested in the new mechanical devices of his day, such as fountains with moving parts, and wind-up mechanical models of birds and other animals. To him, living animals were simply more elaborate versions of these models, whereas humans alone could think. In response to the religious mores of his day, he assigned souls to humans. Thus, humans were endowed with minds and souls. *Cogito, ergo sum*, 'I think, therefore I am', means that of all life on this planet only humans are beings. This constructed divide between humans and other animals, which we call the Cartesian model, still guides our attitudes today, despite the advent of Darwin's theory of evolution. To put humans at an insuperable distance from the animal world was, of course, consistent with the Judaeo-Christian biblical story of creation. 'Man' was placed at the pinnacle of creation, destined to rule over nature and justified in using it to serve 'his' own ends.

In 1859, Charles Darwin wrote *The Origin of the Species* and with it he opened up the great debate about evolution. Continuity of species, changing from one to the next by the process of natural selection, was the central premise of this theory. Most of us know of Darwin's theory about evolution through natural selection of physical characteristics. Characteristics that enhance survival and reproduction of a species in its particular environment are retained and the others are lost. Darwin was also interested in the evolution of behaviour and of the mind. He wrote about this in his book *The Expression of the Emotions in Man and Animals* published in 1872. To Darwin and many of his colleagues (e.g. George Romanes who wrote *Animal Intelligence* in 1882) continuity of species development implied a gradual evolution of mental capabilities, just as occurs for the physical characteristics of animals.

Thus, in contrast to the dominant Cartesian model of the time, Darwin outlined a theory for gradually increasing complexity of mental abilities across species, rather than a sudden appearance of consciousness and awareness in humans. This aspect of Darwin's theory has been largely ignored, even by the majority of scientists who accept his theory of evolution for physical characteristics. In fact, the evolution of the mind has been a rather taboo topic for scientists.

Traditionally, scientists who study the behaviour of animals (i.e. ethologists, comparative psychologists, psychobiologists and others) have strenuously avoided attributing consciousness to animals. Attributing 'human'-like characteristics to animals, known as anthropomorphism, has been frowned upon by scientists. Despite the rise of the sciences that focus on higher processing in the brain (i.e. more complex processing, referred to as cognition) and on complex behaviours performed by animals, it remains decidedly suspect for 'good' scientists to enter into discussions about whether animals have thoughts or feelings. From a scientific position, it is considered to be preferable to describe the behaviour in simple stimulus–response terms without reference to thoughts or emotions. Following this behaviouristic approach, it is considered scientifically unsound to even contemplate whether animals think.

Avoidance of anthropomorphism is also in tune with the predominant cultural and religious attitudes of the Western world, and this makes it clear why so few have contested the absolute validity of the anti-anthropomorphic, scientific position. Most ethologists (scientists who study animal behaviour in the field or laboratory) adopt the position that nature selects for apparently purposeful behaviour in animals, but the animals themselves are not considered to be conscious of the reasons why they decide to behave in particular ways. By purposeful behaviour scientists mean behaviour that ensures the survival of the species. In other words, if the behaviour makes sense to us, from our vantage point, we view it as purposeful. Animals may behave in ways that seem to predict future events but most ethologists claim that only the human observer might be aware of any purpose in these behaviours. Animals are seen to choose between alternatives but it is not believed that they weigh up the alternatives, think about them and then decide. Animals are said to form 'search images', or even 'internal representations', but they are not thought to have *ideas*. This parsimonious approach typifies the scientific study of animal behaviour and it has been useful for describing many aspects of behaviour, providing tangible explanations without alluding to the intangibles of thought processes.

It is possible to describe a great deal of behaviour, of humans as well as animals, without reference to any underlying thought processes. Indeed, some scientists who study animal behaviour claim that this approach is essential for rigorous investigation of behaviour. Undeniably, to adopt such a limited approach to the study of human behaviour would leave out the most important aspects of our species. It follows, therefore, that scientific approaches that categorically deny the possibility that animals may be conscious must, ultimately, limit our understanding of the behaviour of animals. In the past, to even raise the question of consciousness in animals exposed a scientist to ridicule. Nevertheless, for the first time in many decades, some scientists are now beginning to address the issue of consciousness in animals systematically. This new move was largely precipitated by the ethologist Donald Griffin, who wrote the book *Animal Thinking*, published in 1984. I can remember what a stir he caused at the International Ethological Conference held at Oxford University in 1981 when he first addressed the idea of consciousness in animals. The audience was certainly not with him then, but now more ethologists, as well as scientists in some other disciplines, are taking part in these discussions.

Ironically, we are doing so at a time when more and more species of animals are becoming extinct as a result of human intervention. It is paramount in my mind that we are at the brink of driving our nearest relatives, the great apes, to extinction by destruction of their habitats. A pending loss of such magnitude must give impetus to the debate about consciousness in animals.

Research of consciousness in animals is made especially difficult by our inability to use language to communicate with them. Language is the main means by which we know whether another human is conscious. Another person can tell you what he or she is thinking about but an animal cannot, or at least we cannot understand what it is communicating. Without the ability to communicate with animals by using language, we may be unable to access thought processes that might be conscious. As Andrew Whiten says in the beginning of his book *Natural Theories of the Mind*, 'How can we read minds when we see only behaviour?'.

Some scholars argue that language is an essential prerequisite for consciousness. They also believe that consciousness can be revealed only by the use of language. Thus, the reasoning is circular. If you want to limit consciousness to language communication, by definition animals will not have it, unless they can learn human language. That is exactly what some apes have done. Humans have taught some chimpanzees, orang-utans and gorillas to communicate in English by using sign language or symbols. The choice to use sign language or symbols rather than spoken language was made because the structure of the vocal apparatus of apes does not allow them to make human speech sounds. The chimpanzee Washoe was the first to be taught to use the human communication system. In the 1960s she was taught to use Ameslan, American Sign Language, by Beatrix and Allen Gardner at the University of Nevada, USA. Another chimpanzee followed soon after: Sarah, who was taught by David Premack of the University of Pennsylvania, USA, to use symbols for words. She was given coloured plastic shapes backed with metal and was able to communicate by making them adhere to a magnetised board instead of using gestures. As we will see in chapters 3 and 6, using signs or symbols apes can communicate about objects and events not in their immediate environment.

By teaching apes to communicate with us, we open up one channel by which we might determine whether consciousness exists, but I would like to say from the outset that I do not adhere to the notion that consciousness can be expressed only by use of language and I do not believe that we should use language as a barrier to investigating consciousness in nonhuman animals. We do not say that humans who have lost the ability to use language lack consciousness. For example, a person who has suffered a stroke that has destroyed the centres of the brain used for control of speech and analysis of language, usually in the left hemisphere, is not considered to have lost the ability of consciousness or self-awareness, and rightly so. Why then should an animal that does not communicate by using human language be assumed to lack consciousness?

There is another twist to this perspective. Is language unique to humans? Perhaps the vocalisations of animals have much in common with human language. The complexity of song in birds might be suggestive of this. In some species, forms of communication other than vocalisations are used to communicate and these might serve as a 'language', even though they may not have all of the same characteristics as human language. For example, facial expressions, body posture and even odours may be used to transmit information from one individual to another. The question is, do any of the many and varied forms of communication that animals use have anything in common with human language and are they used to communicate about events that have occurred in the past or in another place or to make plans for the future? Communication in animals is a topic for another book, indeed the next one that I am writing with my colleague, Gisela Kaplan, but here I just want to draw attention to the fact that we might also debate the continuity of language across evolutionary time versus the discontinuous appearance of language in humans. To find out, from a human-centred position, one might ask, 'Do animals have the mental capacity for language?'.

At this point we could ask what exactly we mean by 'language' and enter into the controversy that has surrounded the teaching of sign language to apes. This exceptionally heated controversy began in the wake of the research with Washoe. Sceptics, in particular the American psychologist Herb Terrace, argued that certain controls were missing from these studies and that Washoe and Sarah did not use language like humans. From his own work with a sign-language-trained chimpanzee, called Nim, he deduced that what had at first appeared to be self-generated conversation in the chimpanzees was only mimicry, albeit clever mimicry, of subtle signs that the humans were not conscious of sending to the chimpanzees—similar to the case of Clever Hans, the horse that was said to be able to count but was really relying on subtle cues from his trainer (see the book by Robert Boakes for more on Clever Hans). Personally, I believe that Terrace went out of his way to find reasons to criticise and that he failed to understand the bond that must develop between animal and human teacher for

10

communication to occur, even though he trained a chimpanzee himself. Also, following on from their original research with Washoe, the Gardners trained several more chimpanzees and tested their abilities to sign in response to seeing images on a television screen placed in a room without the presence of human observers. Without any cues that might be provided by a human, the chimpanzees were able to sign accurately.

The language-in-apes controversy is still with us today but the recent work of Susan Savage-Rumbaugh, who has taught Kanzi, a pygmy chimpanzee (also called a bonobo), to point at symbols in order to communicate, has quelled at least some of the scepticism. Kanzi has been tested for his ability not just to generate communication using the symbols but also, more importantly, to understand spoken English. Kanzi's ability to understand requests improves when the requests are made in syntactically complete sentences, as compared with truncated, pidgin-English. He has demonstrated the ability to comprehend English, and I would wager that many more animal species might be able to do the same. This might be particularly true of animals that share our homes and so are raised in close contact with human language (see chapter 5). I am suggesting that the ability to comprehend at least some aspects of language may have preceded the ability to speak. In my opinion, it is entirely possible that some of the mental processes that are used for language in humans are present in animals but may be used for other functions, perhaps in part for their own communication systems but also for complex perception, for forming mental representations of the visual world and for problem solving.

I am drawing attention to the possibility of an evolutionary continuity for both language and consciousness, together or separately. Why would these continuities today be more disturbing than the widely accepted continuity of physical (i.e. morphological) characteristics across related species? We recall the enormous controversy that surrounded Charles Darwin's theory for the evolution of morphological characteristics in the last century. I think the reason why consideration of the gradual evolution of language and consciousness is so hotly debated in some circles is that in language and consciousness we have located the essence of what we now consider to be human, and human alone.

It does, of course, remain possible that the brain evolved to reach a level of complexity sufficient for consciousness only in humans. When the brain reached a certain level of complexity there might have been a quantal leap in information processing, and thus consciousness as well as language bloomed *de novo*. However, it is equally possible that we humans are just another step in the continuity of evolution of mind and that, while important, language may not be the only manifestation of or prerequisite for consciousness.

It is also possible that different forms of consciousness may have evolved many times over in different species. Thus, we might expect to find different degrees of consciousness and different manifestations of consciousness in different species. Evolution is not a single, linear trajectory. There are numerous divergences, often pictured as branches of the evolutionary tree. The different routes of evolution occur as the result of adaptations to different environments. Thus, species may be just as complex as each other, and just as adapted to their own particular environment, but they may also be cognitively very different from each other. Just as different species rely on different senses, some attending to sounds more than vision and others more to smells, so too might their mental processes differ. We must look to studies of animal behaviour to try to answer this.

To try to reconstruct the steps of evolution, we can study only the existing species because behaviour leaves no fossil record. From present-day species we have to deduce the behaviour of their ancestors, and this is the case for humans as well as nonhuman animals. From observation of the behaviour of a species we have to decide how they 'think' or, to use terminology that is more acceptable to scientists, we have to assess their cognitive capacity.

Cognition is the term used to describe the more complex processes that occur in brains, human and animal. It includes higher processing of information, decision making, learning of more complex tasks, problem solving and so on. Complex cognition is frequently considered to mean much the same as intelligence, which we will discuss in more detail in chapter 3. Here it is important to point out that intelligence has many meanings. We sometimes use the term to refer to a characteristic of an individual, sometimes to a characteristic of an entire species and sometimes to a specific behaviour. As David McFarland, Reader in Animal Behaviour at Oxford University, UK, has pointed out, intelligent behaviour can occur without cognitive processes being involved: humans can orient in the environment, using our spatial abilities, in ways that would seem very intelligent if performed by a robot. The control of spatial orientation does not necessarily require cognition, even though the behaviour produced appears to be intelligent. McFarland says that cognitive ability is not merely the ability to produce clever behaviour. Cognition depends not on fixed responses adapted to well-specified situations but on complex processing of new, or less common, information.

Of course, a distinction must be made also between complex cognition and consciousness. It may be possible for complex cognition to occur without consciousness occurring, although it is quite clear that consciousness would not be possible without the ability for complex cognition.

Consciousness is related to awareness, intelligence and complex cognition, as well as language. Consciousness may be manifested in self-awareness; awareness of others; intentional behaviour, including intentional communication; deception of others; and in the ability to make mental and symbolic representations. It is my guess that consciousness will be reflected in an integration of many, if not all, of these behaviours and modes of cognition. The chapters to

13

follow will examine evidence that has bearing on all of these perspectives.

From the beginning, I am acutely aware that the history of ideas has set the starting point with animals behind the line, without consciousness, and my task is to see whether that assumption might be incorrect. In a different world, in a different place or time, I might equally well be starting with the assumption that animals have consciousness, that they are beings. Then my task would be to see whether that may not be so. As a scientist, I would still be faced with weighing up the evidence for and against consciousness but my approach would be somewhat different. I am also conscious of the possibility that to adhere to the belief that no animal has consciousness until it can be proven otherwise may be a justification for exploitation of animals. It is not an exaggeration to say that believing we humans alone possess consciousness has permitted all manner of abuse and exploitation of animals. The debate about consciousness or awareness in animals is central to issues of animal welfare. Although current concerns for the welfare of animals in research and agriculture have focussed on the ability of animals to feel pain, future considerations will have to take into account new findings about awareness and complex cognition in animals.

CHAPTER TWO

AWARENESS OF SELF AND OTHERS

Awareness of self is a central aspect of consciousness. At a basic level, self-awareness means to be aware of one's own feelings or emotions and to be conscious of pain, but self-awareness also includes awareness of one's body (e.g. allowing recognition of oneself in a mirror), one's state of mind, one's self in a social context, and numerous other, ill-defined attributes that we would assign to ourselves.

We have discussed how, in the seventeenth century, Descartes and many others advocated the view that animals were machines, differing from human-made machines only in their degree of complexity. According to the Cartesian view, the yelping of a beaten dog was merely the creaking of the animal's clockwork machinery. Today most people believe that all vertebrate animals, at least, can feel pain. Whether the more primitive species, the invertebrates (animals without backbones, such as jelly fish and insects), can feel pain remains unresolved and usually ignored. Some animal species may react to a painful stimulus by withdrawing from it without being conscious of that stimulus and without feeling pain. I would be most surprised, however, if all invertebrates were completely unable to feel pain. The acquisition of a backbone signifies an important step in evolution, and many other characteristics were acquired with it, but many invertebrates have quite complex nervous systems and perform remarkably complex behaviours.

The ability of vertebrate animals to feel physical pain has been the main concern of those interested in animal welfare. As a consequence of accepting that vertebrate animals feel pain, most Western countries have introduced legislation to protect vertebrate animals used in research. However, few people would extend this line of thinking to consider that animals may feel pleasure, happiness, love, hate and mental pain. We seem to want to reserve these emotions and other higher aspects of feeling for humans, but are we correct in doing so? Perhaps awareness of pain was the first aspect of sentience (i.e. conscious experience) to evolve and then awareness of emotional feelings was the next step. It is this next step that most of us are reluctant to grant to animals.

Yet, by their facial expressions, body postures and vocalisations animals may express distress and pleasure. For example, a young chick emits loud calls with a descending pitch when it is distressed by separation from the hen or by being cold, and it emits softer calls of ascending pitch when the hen returns or when feeding. The hen can interpret these calls and respond accordingly. But does the chick actually *feel* distressed or unhappiness when the hen leaves and pleasure when she returns? Most, if not all, species of animals can express behavioural states of various emotions but are they aware that they are doing so and can they reflect on these feelings?

Developing an awareness of self

We know that in humans awareness of self goes well beyond feeling and expressing emotions. Humans develop a sense of self by the accumulation of experiences, and to do this we rely on memories of those experiences. As far as we can assess, we begin life without a well-developed sense of self, if we have one at all. The new-born baby can react to stimulation from the environment. Indeed, the baby's first expression of feeling is to cry, perhaps to express pain. But we cannot remember if we felt pain or any other feeling at the beginning of our lives. The ability to be aware of the self appears to develop with age or, at least, the ability to form a memory of it does so.

A human infant is at first unable to perceive itself as separate from its surrounding environment. That environment includes other individuals, particularly the mother, as well as the physical environment. In time the infant learns that it cannot actually grasp attractive objects out of its reach and that its feet are part of the self. The developing brain of the infant forms maps of the infant's own body and of the world around it. Animals do likewise, and neurophysiologists recording from nerve cells in the brain (neurons) have found such maps laid out in different regions of the cortex of cats and monkeys, the only species that have been studied in detail.

We do not know with complete accuracy when human infants become aware of their own feelings and when they begin to develop a sense of self. To discover this with absolute certainty we would need to communicate with them and this cannot be done until their ability to use language has developed sufficiently to tell us what and how they feel. The problem is exactly the same as it is for animals. To find out the cognitive processes of human infants before they can speak, we are limited to the same techniques that must be used for animals. Yet we attribute awareness of emotions to human infants before they can speak even though most of us do not do so for animals.

Most of the psychological evidence indicates that human infants develop a concept of the self from around twelve to twenty-four months of age. At around twelve months of age, the infant will look to where another person is looking or pointing, a behaviour referred to as 'joint attention' and marking the beginning of a concept of self, as well as a concept of others. By eighteen to twenty-four months infants can recognise themselves in mirrors, meaning that they are aware of their own physical attributes. Awareness of self and others continues to develop, and between the ages of three and five years humans develop the ability to understand the notion of a false belief. A child of about this age can also attribute different mental states to other people. For example, a four-year-old child who sees another person peering into a box can understand that that person knows the contents of the box, whereas another person who did not look in the box does not know. Thus, if the second person volunteers information about the contents of the box, the four-year-old child knows that it is false information. Tests of awareness such as these are, inevitably, confounded by the language development of the child and thus it may not be coincidental that the age of attaining mental attribution is from three to five years. Even in tasks that do not require a response in language, communication between the experimenter and subject may be confounded by the level of language acquisition.

Let us return to an earlier state of development, well before that at which a human may be acquiring language. One of the principles of development of the sensory systems (sight, hearing, touch, taste and smell) is that they come into function sequentially. For example, the ability to respond to sensory stimulation begins with touch and taste and then progresses to hearing and finally vision. This pattern is conserved across almost all vertebrate species and it has been much studied. The sense of smell usually begins early but it varies between species. Self-awareness in humans develops sequentially also, at least in its early stages, but we know relatively little about that process. The development of self-awareness of feelings and emotions possibly begins with the perception and awareness of pain and hunger, followed by awareness of discontent and pleasure, developing to love and hate and so on. Perhaps animals get so far along this sequence of development of self-awareness and stop before it is completed, the stopping point depending on the species. Species that evolved earlier may stop developing awareness at an earlier stage compared with more complex, later evolving species.

A notion such as this is a very old one. It is referred to as recapitulation, as it assumes that development recapitulates evolution. Originally it was applied to the development of physical characteristics. For example, the development of the human foetus through stages with gills and a tail and with webbed fingers is said to reflect our evolutionary origins from fish and amphibia. If this is the case for physical characteristics, it might be true also for the cognitive processes that underlie the development of self-awareness.

During the early foetal stages of development, the human foetus may respond to touch by moving but it is most unlikely to be aware of doing so. At this stage of development it may resemble a lower, invertebrate species. At later stages of gestation, the human foetus responds to pain-inducing stimuli and it may be able to feel pain, although it may not yet feel emotions. At this stage it might be like a slightly more highly evolved species but perhaps not yet a vertebrate species. Eventually, emotional feeling and self-awareness will develop, after birth.

To consider that development reflects evolution does not mean that the development of self-awareness is controlled by an inherited program (i.e. by the genes). In fact, learning and memory formation are essential to the development of self-awareness. Experience provides the building blocks for the self. The human individual emerges as a result of the ability to feel and to store memories that can be recalled and applied in new situations and contexts. We learn to be 'us' and the end result of this is a unique human being. We are not clones of each other although we may have some things in common with others. Self-awareness is being conscious of both the differences and similarities between one's self and others. We learn to recognise ourselves both as physical entities (e.g. when we look into a mirror) and as mental entities. We are able to reflect on ourselves and we rely on our memories to do so.

Each animal is an individual. Within a species individuals vary in their abilities to learn, to take the lead in different situations and to solve problems, in their reactions to novel situations and in their activity levels, to name just a few of the potential sources of difference. These differences between individuals may depend on temperament, perhaps in part inherited but also moulded by experiences beginning even before birth. Each individual animal has different experiences and thus forms different memories that are built up over a life-time, just as in humans. Temperament itself is moulded by experience. In other words, the uniqueness of an individual is not simply encoded in the enormous diversity of our genetic code (our inheritance) but is established by our unique experiences encoded in our memories. It is the collection of memories that becomes part of the self. Thus, the complexity of an individual self must depend, in part at least, on the number and variety of memories that have been formed. Of course, the individual might not be aware of some, or even any, of the memories that it has formed. Cockroaches can learn and form memories but are not likely to have self-awareness. Where there is self-awareness, however, the complexity of that self-awareness depends on the memories of which the individual is aware.

Species with more complex nervous systems may form more detailed memories and use more complicated communication systems. Even the young domestic chick has at least fifteen different recognisable calls. Also, the chick possesses one of the characteristics essential for being an individual. It can acquire information and encode memories. These stored memories guide its future behaviour. In fact, we know that a chick can make memories even before it hatches. It hears the hen's vocalisations when it is still an embryo inside the egg and learns their characteristics. This is also known to occur in duck embryos and even in lambs before birth. Learning and making memories before hatching or birth is probably characteristic of all precocial species, ones in which the young are born in a relatively well-developed state, but it may also occur in species that are not precocial.

After hatching, the chick learns rapidly about the visual characteristics of the hen (referred to as imprinting) and, in doing so, it forms an attachment to her. This attachment ensures that the chick follows the hen as she moves away from the nest. The chick also learns to recognise its siblings and, as early as three days after hatching, it can recognise the familiar chicks from unfamiliar chicks. If a chick is put into the centre of an alley way with a familiar cagemate behind a transparent plastic partition at one end and an unfamiliar chick behind a similar partition at the other end, it will make a choice and approach the familiar chick (Fig. 2.1). This means that the chick can distinguish one chick from another and that it can recognise that one of the chicks is familiar, that it matches its memory of that chick.

These are remarkable abilities for a young animal but, although recognition of other individuals is a prerequisite for awareness of others, it does not, necessarily, indicate that the chick is aware of itself. Some people believe that

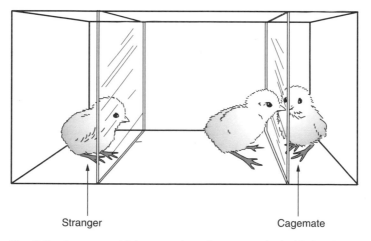

Stranger   Cagemate

Fig. 2.1 A young chick recognises its cagemate behind a transparent panel and approaches it. A stranger is not approached
Source: Modified from Vallortigara and Andrew, 1991.

21

the chick behaves like a custom-designed machine shaped, or adapted, by its own individual environment. There is no way of disproving this mechanistic concept with presently available evidence but it is still apparent that the young chick is a much more complex creature than we used to think. More examples of the chick's complexity of behaviour will be given later.

The development of self-awareness may be dependent on the social environment in which an animal is raised, as well as on age and other individual characteristics. For example, the gorilla Koko, raised by humans, showed recognition of herself in mirrors by the time she was about fours years old, whereas some other gorillas raised with less contact with humans have failed to do so.

Recognition of one's image in a mirror is used as a measure of self-recognition, as I will discuss in the next section. If this behaviour indicates self-awareness, and there is considerable debate about whether it does, it is but one aspect of self-awareness. There must be many and various forms of self-awareness, and not all individuals or all species are likely to show every form of self-awareness. Indeed, the self is a rather elusive thing, not easily tied down to a simple measure, if it can be at all. The psychologist William James, writing in the early part of this century, divided the self into three parts: the 'material' self, which takes into account only the physical aspects of the body; the 'spiritual' self, referring to beliefs about one's moral standing and future directions and hopes; and the 'social' self, one's concept of self as it might be regarded by others. To these 'selves' I would add the self that has knowledge of one's own past and of one's motives and desires. Without entering into discussion on the likely validity and relative contributions of these aspects of self, it is obvious that the self of humans has many different facets, of which some may be linked to each other and others may be quite separate. The same is likely to be true of the self of animals.

Self-recognition in mirrors

When an animal looks in the mirror does it know that it is seeing itself? Recognition of self in a mirror image has received much attention as an experimental method of assessing self-awareness in animals. In my opinion, there has been too much weight placed on a limited number of quite inadequately controlled experiments with mirrors. The manner in which members of different species behave when they see their images in mirrors is fascinating in its own right and, whether the individual responds to the image as if it were another member of its species or itself, does tell us something about self-recognition—but a specific type of self-recognition based on the visual representation of self in a left/right inverted image that moves when the individual moves. It does not provide information about recognition of self using auditory, olfactory or tactile information, all of which are important aspects of the self-image, and it certainly tells us little, if anything, of the mental aspects of self, although researchers who have used the mirror technique have often led us to believe that they are studying self in a more total sense than is actually the case. This is why I say that the research on self-recognition in mirrors has assumed rather too central a place in the question of self-recognition in animals.

When most animals first see their images in mirrors they treat them as though they were another member of their own species. They may attack the image, display fear or engage in social behaviours towards it. They may go behind the mirror to see where the rest of the body is, as did my donkey when he once came inside the house and caught sight of himself in a hallway mirror. Most species do not recognise that the image is of themselves even after prolonged exposure to it. This is, apparently, not the case for chimpanzees. After five to thirty minutes exposure to a mirror, chimpanzees begin to indulge in self-exploratory behaviours using the mirror. They may use the mirror image to see parts of their bodies that they cannot see directly. They protrude the tongue, clean the teeth or nose and inspect their genitalia. Much of the behaviour in front of a mirror is playful. For example, one chimpanzee stuck celery leaves up her nose and hit at them with her fingers. All of these chimpanzees appear to have recognised that the image is of self. Nevertheless, although performance of these behaviours in front of the mirror does not appear to be coincidental, more rigorous tests are necessary to prove this.

In the 1970s Gordon Gallup of the State University of New York, USA, attempted to see if a chimpanzee could recognise itself in a mirror by putting a spot of red dye on the chimpanzee's forehead and then waiting to see whether the chimpanzee touched the spot on the image in the mirror first, indicating that it did not see the image as self, or whether it immediately touched the spot on its own forehead. Gallup tested four chimpanzees, born in the wild, captured and brought to his laboratory in the United States. Prior to the experiment, they had had little or no experience with mirrors. At the commencement of the experiment, each was caged in a separate, small cage and a full-length mirror was placed in front of the cage. The behaviour of each chimpanzee at the mirror could be observed through a peep hole in the wall. At first the chimpanzees treated their image as if it were another chimpanzee, and they engaged in head bobbing and vocalising and threatened the image. But, after about three days, they began to perform self-directed behaviour, using the mirror to groom parts of the body that they could not see without the mirror, making faces at the mirror, blowing bubbles and manipulating wads of food in their lips while looking in the mirror (Fig. 2.2). It appeared that they had learnt to recognise themselves in the mirror. Then, after they had ten days of exposure to the mirror, Gallup anaesthetised each chimpanzee and, when it was unconscious, applied a spot of red dye to the forehead and tip of one ear. The chimpanzee was returned to its cage without the mirror being present. Four hours later, by which time Gallup claimed they had recovered

Fig. 2.2 A chimpanzee recognises her image in a mirror and examines parts of her body that cannot be seen directly *Source:* Adapted from Povinelli and Preuss, 1995.

from the anaesthetic, the number of times that they touched the spots of dye was recorded over a thirty minute interval. They did not touch the dye very often. Then the mirror was returned to the front of the cage and the same behaviour was scored again. Now there was a several-fold increase in the number of times that the chimpanzees

25

touched the red spots on their own foreheads or ears while looking in the mirror. Gallup concluded that this showed they were able to recognise themselves in the mirror and were therefore self-aware.

While this result was exciting enough at the time to be published in *Science*, one of the leading scientific publications for newsworthy information, it has subsequently been criticised, particularly by Celia Heyes of University College London, UK. First, there was no control for the effects of the anaesthetic. Just four hours after being anaesthetised the chimpanzees might be first less active and then more active as the anaesthetic wears off. In other words, this could have confounded the results that Gallup collected. Although Gallup also tested two other wild-born chimpanzees that had no experience with mirrors, he did not have an exact control in which he repeated the entire experiment but simply applied a colourless dye to the forehead and ear instead of the red dye. The two chimpanzees that had no prior experience with mirrors did not show increased touching of the red spot when they were tested in front of the mirror. This could have been because they had to learn to recognise themselves in the mirror, as Gallup suggested, but it could have been caused by a number of other factors related to being more stressed or, perhaps, being less interested in the task in a general sense.

In a later experiment Gallup did apply red dye to the wrist and stomach, parts of the body that could be seen without the aid of a mirror, and the amount that the chimpanzees touched these did not increase in front of the mirror. In other words, the increased touching of the marked forehead and ear is specific and not a general increase in touching that might be an after-effect of the anaesthetic. However, it would have been preferable to allow the chimpanzees to recover until at least the next day before scoring their behaviour with and without the mirror. Anaesthetics can have very long-lasting effects and result in quite unexpected behaviours.

In response to Heyes' criticism of these experimental

methods, an experiment was conducted in which one chimpanzee had a spot of red dye placed on her right eyebrow and another placed on her left ear. The amount of touching of both eyes and ears, marked and unmarked, was scored. With the mirror present, there was increased touching of the marked eyebrow and ear but not of the unmarked one. Thus, the response is specific for the marked skin only, but so far only one chimpanzee has been tested in this way.

Gallup also tested some macaque monkeys using the same procedure that he had used with the chimpanzees and they persisted in reacting to the image in the mirror as if it were another monkey. They showed no decline in directing social behaviour to the monkey in the mirror even after more than two hundred hours of exposure. Gallup concluded that there is a 'qualitative psychological difference' between chimpanzees and monkeys and that the capacity for self-recognition may 'not extend below' humans and the great apes. Contrary to earlier belief, humans are not alone in mirror image recognition but, according to Gallup, we are in a select group together with the great apes and different from all other species. We will see later that this conclusion is incorrect.

Other researchers have found the same results as Gallup using the red-spot test with orang-utans and gorillas, although Gallup himself was unable to get gorillas to respond to self in the mirror. The gorilla Koko, however, who uses sign language, does respond to mirrors in the same way that chimpanzees do, and the same has been found in two other gorillas that have been taught to use sign language. In fact, Koko used the mirror to alter her appearance: she made up her face with chalk and scrutinised the result in the mirror.

It should be mentioned that another researcher applied dye to the forehead of a chimpanzee when it was asleep and, after waking, it showed no increase in touching the spot when in front of a mirror. Perhaps the anaesthetic had caused a misleading result in Gallup's experiments, but individuals can differ and only one chimpanzee was tested by applying the dye during its sleep. In fact, researchers at another laboratory attempted to repeat Gallup's mark test using eleven chimpanzees and applying the dye when they were anaesthetised. In this study only one of the chimpanzees displayed clear self-directed behaviour in response to seeing the mark on her forehead. The researchers suggested that individual differences might explain why they found this result, but differences in methodology could also explain why only one of their chimpanzees performed the same as those tested by Gallup. In fact, they began testing the chimpanzees only two-and-a-half to three hours after the anaesthetic, and this could have been a problem. The chimpanzees might have been too drowsy at the time they were tested, or they might have felt ill. Also, the anaesthetic used was different from that used by Gallup and it may have lasted for a different time or had different after-effects.

So far there has been no completely convincing experiment with sufficient subjects and controls to permit a definite conclusion to be reached about self-recognition in mirrors by chimpanzees or any other species. However, I must say that the published photographs of chimpanzees performing in front of a mirror (see those in the book by Richard Byrne, *The Thinking Ape*, or in the book by Sue Taylor Parker and colleagues, *Self-awareness in Animals and Humans*), protruding the tongue, and so on, give a clear impression that they are recognising themselves. Nevertheless, we must await rigorously controlled experiments to be absolutely sure.

The apparent absence of ability in monkeys, as opposed to apes, to respond to their image as self may have been merely a result of not using an appropriate method for testing them. Marc Hauser and colleagues at Harvard University, USA, chose to test cotton-top tamarins (monkeys from South America) with mirrors, and to make sure that they would attend to the spot marked with dye they applied differently coloured dyes to the mane of hair on top of the monkey's heads. This is a visually distinctive feature of the species and one likely to be used in social situations. The tamarins with colour-dyed hair looked in the mirror longer than control tamarins that had only white dye applied on their cotton tops. By including this control group, Hauser eliminated the possibility that the after-effect of anaesthetic could explain the results, but looking for longer in the mirror could have had something to do with being attracted by the colour of the colour-dyed hair rather than recognition of self. However, only the individuals with colour-dyed hair, and prior experience with mirrors, touched their heads while looking in the mirror and, in addition, some of the monkeys used the mirror to examine inaccessible parts of their own bodies, as the chimpanzees had done. Thus, this species of monkey, at least, shows some sort of mirror self-recognition. Species may vary in what parts of the body they attend to, and the dye should be placed on these parts. Species also vary in the amount of social behaviour that they display and this might be another factor in the mirror test, since attention to the image involves social behaviour.

The need for considering species differences in mirror recognition tests is highlighted by a study of this behaviour in elephants conducted by Daniel Povinelli. Two Asian elephants at the National Zoological Park in Washington, USA, were tested with a mirror measuring 105 x 241 cm. This is a large mirror—but not compared with an elephant. We must also take into account that an elephant's eye is on the side of the head. Elephants have some frontal vision, but mainly they look sideways. Therefore, they may recognise each other from the side and perhaps the whole side, not just the head. The entire side of an elephant was not always visible in the mirror. Added to this, elephants may rely on vocalisations, odours and tactile sensations to recognise self and others. They would receive none of these cues from their images in the mirror. In fact the elephants paid little attention to their images in the mirror and, therefore, Povinelli concluded that they fail to show self-recognition.

Certainly, they may fail to recognise themselves using visual cues alone, but this experiment tells us nothing more than that. Better designed experiments are required.

If mirror recognition does occur in animals, what does it tell us about self-awareness? Does mirror recognition reflect superior cognitive abilities? A paper by Epstein and others at Harvard University, USA, reported that pigeons can use a mirror to locate a coloured spot placed on the breast and hidden from direct view by a bib around the neck. Each pigeon was first trained to peck at blue spots elsewhere on its body by rewarding it with food each time it pecked at a spot on the wing, abdomen and so on. They were also rewarded for pecking at blue dots in the cage. Finally, the blue dot was located under the bib where it was visible only by using a mirror. The pigeons saw the dot in the mirror but, rather than pecking the image in the mirror, they bent the head down to attempt to peck at the spot under the bib. The pigeons reacted to the mirror in the same way as had Gallup's chimpanzees. Instead of concluding that pigeons may be as intelligent as chimpanzees or, at least, that they might have an equivalent ability to recognise self in the mirror, the researchers said 'Although similar behaviour in primates has been attributed to a self-concept or other cognitive process, the present example suggests an account in terms of environmental events'. The assumption they made was that, if a bird can do it, it cannot be complex behaviour and it cannot indicate self-awareness of any sort. We now know that pigeons are capable of complex behaviours that rival those of primates, and this will be discussed further in chapter 3.

One of the most important distinctions to be made about the behaviour of animals towards their reflections in mirrors is whether they are showing social behaviour because they see the image as another member of their species or whether they are examining themselves. As social behaviour varies considerably between species, each species has to be considered on its own terms. Some species are more sociable than others, and so are some individuals.

Also, the kind of behaviours that are used socially varies. Ken Marten and Suchi Psarakos, in Hawaii, USA, have tackled this problem in dolphins by looking at their behaviour towards mirrors and video-images and comparing them with social behaviour involving real dolphins. They were able to conclude that self-examination behaviour, as opposed to social behaviour, did occur in the mirror and video-image situations. In addition, they carried out the dye marking test, but used zinc cream instead of red dye. The dolphins appeared to be examining the marked areas of their bodies in the mirror and the results suggested that they were able to recognise themselves.

In time it is most likely that well-designed experiments will demonstrate that many species can recognise themselves in mirrors, and also in photographs and video playback sequences. We might also discover that recognition of the physical self is not confined to the visual image, and that some species are more dependent on their own vocalisations, odour or tactile sensations in order to recognise self. While mirror self-recognition remains interesting, we should be wary of reading too much into it. The concept of self-awareness encompasses much more than one's physical attributes. As I have said previously, mental attributes are a part of the self not reflected in mirrors. Self-recognition in mirrors, photographs or on film is only one small facet of self-awareness.

Awareness of others

All animals interact with each other to varying degrees and at different times in their lives. They communicate with each other by making vocalisations, by displaying their plumage or moving in particular ways, or by emitting odours or other signals. But these kinds of social behaviour may not involve awareness of others as separate selves, as it were. The fact that a species has social behaviour does not tell us that the members of that species are consciously aware of the physical, mental or emotional states of others.

31

A pet dog that becomes miserable when its owner is sad or ill may be aware of its owner's state of mind and emotions or it may be merely mimicking the owner's behaviour. However, if the dog acts on the information about its owner's mind state in a meaningful way, that would tell us something different. A classic case of the latter might be the dog that runs to get help when its owner is in trouble and then leads the helper to its owner. There may be a more trivial explanation for the dog's behaviour than it being consciously aware of the state of the owner, but let us consider another example that Marian Dawkins has outlined in her book *Through Our Eyes Only?* (cited as further reading for chapter 1). Rats are social animals and they can learn to avoid poison baits by observing the behaviour of a companion that has been made ill by consuming a bait. Bennet Galef of McMaster University, Canada, raised rats in pairs and then took one member of the pair away to feed it a novel food. After the rat was returned with the smell of the new food on its breath, its companion would follow suit and readily eat the same food when given a choice between it and another novel food. So far, the companion may only have mimicked its partner but, if Galef made the first rat ill after it had fed on the same novel food and returned it to its companion when it was feeling ill, the companion would reject that food when it was offered. In other words, the companion had assessed the physical state of the sick rat and acted on that information. Moreover, the social life of the rat is such that this learnt avoidance of the particular food may be passed on to subsequent generations. A 'cultural' tradition had been established. Through the first rat perceiving and responding to some aspect of its companion's state of health, an important tradition had been acquired by the species. Of course, it might be possible that the rat simply learns to associate the odour of the food with some sort of negative cue from the body posture of the sick rat (i.e. it sees it as a sort of punishment and so gets conditioned not to take the food) but, equally, it might be aware of the other rat's state of health. Once the initial learning has occurred, the actual information that has to be learnt in order to establish the social tradition of food avoidance may not be particularly complex.

Awareness of others may also entail knowing their social status and their relationship to others. An excellent example of the latter comes from the research of Dorothy Cheney and Robert Seyfarth, who have studied the behaviour of wild vervet monkeys in Africa. By recording the vervet monkeys' calls and playing them back to the monkeys at their study site in the field, they were able to assess how the monkeys interpret the calls. Beginning with the observation that mother vervet monkeys run to help their offspring when they scream during rough play, Cheney and Seyfarth designed an experiment that would show whether mothers recognise their own offspring's call when it is played back and whether other, nonrelated females ignore that call. Not only did the nonrelative females ignore the scream for help by not approaching the loud-speaker, whereas the mother approached it, but they also turned to look at the offspring's mother when they heard the scream. That is, they recognised that the scream belonged to the offspring of that particular mother. Thus, vervet monkeys must have a concept of relationships between other members of their group. This ability to recognise relationships may be a basis for being aware of the mental states of others but it is not proof that it occurs. Although monkeys may know the relationships between other members of their group, they may not be able to distinguish between their own state of mind and that of others.

Following the direction of gaze of others and imitation

As mentioned earlier, at around one year old the human infant will follow the direction of gaze of another person and therefore look at the same thing, or at least in the same vicinity, as that person is looking. This behaviour is said to be a prerequisite for being aware of others. Some researchers claim that autistic children do not show following of the direction of eye gaze, consistent with their less-developed awareness of the mental states of others. There may be an aspect of imitation in eye-gaze following, and autistic children are less inclined to imitate. They are less likely to play games that involve imitation of the actions of another human than are normal children. Incidentally, autistic children can recognise themselves in mirrors.

There has been very little investigation into patterns of eye-gaze in animals, despite the fact that eyes and eye patterns are known to be very potent visual signals in animals as diverse as insects, birds and monkeys. However, there have been a few reports of eye-gaze following in apes and monkeys and these suggest that the apes (chimpanzees, orang-utans and gorillas) follow the direction of eye gaze of humans, whereas monkeys do not do so. However, there has not been a sufficient number of controlled studies of this behaviour for these indicators to be accepted as conclusive. Also, it would be more important to know whether there is eye-gaze following of other members of the same species. Apes may follow the direction of eye gaze of their human carers, but would they do likewise for humans with whom they are unfamiliar? Perhaps the monkeys had not formed such a strong bond with their human carers as had the apes, and this is why they did not follow the direction of their carer's eye gaze. There are many controls that need to be performed before we will be able to draw conclusions.

Mutual looking in the same direction is observed commonly in the wild in a wide range of species, but this may simply occur because all members of the group have spotted the same visual stimulus or heard a sound coming from that direction. To be considered as gaze following, one individual must follow the gaze of another simply because that individual is looking there and not because any other cue has been received by the follower. Researchers working on wild primates have reported examples that might meet this requirement but it is difficult to prove that there has been no other signal to cue in the same behaviour. Richard Byrne relates an example that seems convincing. He saw a wild baboon about to be chased by another. The one about to be chased stood on his hind legs in a posture that baboons usually adopt when they have seen a predator in the distance or another troop of baboons in the long grass, and looked intently in one direction. His pursuer stopped the chase and looked in the same direction. No predator was in sight. Byrne assumed that the baboon being chased had used this as a tactic to distract the other one's attention. Such potential deception will be discussed later.

Following the direction of eye gaze would, of course, be most usefully applied to detecting predators. Thus, if one member of a group has detected a predator, the other might follow its direction of gaze to do likewise. This would be strong reason for the evolution of the behaviour, but exactly whether eye-gaze following reflects any aspect of awareness of others could be debated. Relatively straightforward computations might be used to follow another's direction of gaze and these might not necessitate adoption of the other's perspective.

There are many other kinds of imitation behaviour. For example, humans imitate the way in which other people move, perform certain acts, speak and so on. Much of our cultural learning occurs by imitation. Some psychologists claim that imitation is unique to humans and that it is intimately related to self-awareness and being able to take the visual perspective of others. There is, however, convincing evidence that the great apes can imitate. Chimpanzees raised by humans frequently imitate their behaviour and those that have been taught sign language often imitate the signing of humans or other chimpanzees. Anne Russon of Glendon College in Canada[^] has been studying imitation in orang-utans at a rehabilitation centre in Borneo and she has reported that they frequently imitate the behaviour of humans working at the centre. They imitate the gardener by chopping weeds at the edge of the path and collecting them into rows, sweep the floor with a broom, hammer planks together, saw beams of wood, chop wood with a hatchet, use a shovel to dig, and attempt to start a fire using fuel and fanning with a lid, to list but a few of the imitation behaviours that she has observed.

We might now ask what is the difference between imitation and mimicry. Probably the best examples of mimicry in animals can be found amongst those species of bird that perform vocal mimicry. The Australian lyre bird has a remarkable ability to mimic the calls of other species of birds or of nonavian species in its environment, such as the barking of dogs and, as I have heard in Sherbrooke Forest outside of Melbourne, Australia, the garbled speech sounds of a group of humans. They also mimic inanimate sounds, such as passing trains and whistles. These mimicked sounds are incorporated into their song. The same is true of magpies, particularly those raised in close contact with humans, and we are all very familiar with the mimicking of human speech by parrots and cockatoos. Why do we refer to this form of copying behaviour as mimicry and the copying behaviours of humans and other apes as imitation? The latter is considered to involve higher cognition and to be an aspect of consciousness, whereas mimicry is thought to be occurring automatically without self-awareness. But how do we make this distinction in real terms? As a human baby develops awareness of itself does it shift from mimicry to imitation? Very young babies copy the smile of adults, particularly the mother, and we call this imitation, but perhaps it is really mimicry. On the other hand, we do not know that lyre birds, magpies and parrots are using lesser cognitive processes when they copy sounds, particularly during the learning phase when they are acquiring the ability to do so. There must be different forms of copying behaviour, some better referred to as imitation and others as mimicry, but the present use of these separate terms is in reality determined by the attitudes of avian biologists versus primatologists and by our expectations of the species in question.

Awareness and communication

Awareness of self and others may also be a part of communication. When a young chick is distressed, it emits peep calls that attract the hen. Is the chick aware of the fact that it is sending messages to the hen? It might simply produce calls as a read-out or by-product of internal processes, like a machine, not even being aware of feeling discontent, let alone being aware that it is communicating with the hen. That is, the communication may not be intentional. The same questions may be asked of a human baby when it cries for its mother. Being aware of the vocalisations that we make is something that develops with age. The same may be true for the chick.

There is some evidence, although not complete enough for us to be sure, that adult chickens may communicate intentionally and therefore be aware of the fact that they are communicating. Before discussing this I must say something about the calls that chickens use to communicate.

Peter Marler, at the University of California in Davis, USA, and Chris Evans, now at Macquarie University in Australia, discovered that roosters emit alarm calls when they see a predator, such as a hawk flying overhead or even an image of a hawk on a videoscreen overhead, but do so only when they have an audience of other members of their species. The alarm call made in response to seeing an aerial predator is very different from the call made in response to a predator on the ground, such as a dog or raccoon. The aerial alarm call is a long screech, whereas the ground alarm call is a series of short pulses of sound. The presence or absence of an audience does not influence the ground alarm call. Apparently, as Marler and Evans suggest, the call is as much directed at the predator, in an attempt to scare it away, as to other chickens. The calls have specificity that can be interpreted by other members of the species and, indeed, chickens respond appropriately by crouching and looking up when they hear the aerial alarm call or by running for cover or strutting when they hear the ground predator alarm call.

The calls of chickens are quite specific, relaying information that can be interpreted correctly by other chickens, and the calls are emitted only in specific contexts. This would suggest that they are not simply produced impulsively and involuntarily, as many people have believed. Also, they are not simply a read-out of the bird's state of emotion emitted in any context, although emotion may still have a role in their production. The question relevant to awareness is whether the chicken making the call knows that it is sending the message. One could argue, as many have, that the chicken is programmed to emit the alarm call only if a conspecific (another chicken) is present and thus there is no intentional communication.

One way of discovering whether animals communicate intentionally is to see whether they use a call with a specific meaning in an unusual context in order to deceive another animal. Gyger and Marler have reported some evidence that chickens might use calls to deceive. In the presence of food, chickens emit a 'food call' and this attracts other chickens to the food site. The researchers reported incidences in which a rooster issued the food call in the absence of food to deceive a hen into approaching. This example will be discussed later in more detail under the topic of deception.

Vervet monkeys (or green monkeys) also give alarm calls when an audience is present. Seyfarth and Cheney have found that vervet monkeys produce different calls in response to seeing different predators, such as leopards, eagles or snakes. The call given when they see a leopard is a barking sound. When they see an eagle they emit a single cough-like sound, and when they see a snake they chutter. Each of these alarm calls elicits the appropriate form of defence by their conspecifics. If one monkey sees a snake and calls 'snake' in vervet-monkeyese, the others stand erect on their hind limbs and peer into the tall grass, whereas the call 'leopard' sends them scurrying up the nearest tree and the 'eagle' call causes them to look up and take cover. The monkeys certainly seem to be responding as if they know the meaning of the calls. To add further weight to this interpretation, the researchers tested the monkeys using a method of habituation–dishabituation that tests speech perception in human infants. They chose two social contact calls, a *wrr* which is given when the monkeys spot another group of monkeys and a *chutter* which is emitted in aggressive encounters between groups. Thus both calls are associated with groups of monkeys even though they sound very different. First, a subject was exposed repeatedly to the *chutter* of another individual until it no longer responded to the call (i.e. it had habituated to this call). Then the *wrr* call of the same individual was played. The test subject did not respond (i.e. it did not dishabituate). It treated both calls as if they were now the same, having something to do with a group of monkeys that could not be seen and which were now being ignored. Given that the *wrr* sounds very different from the *chutter*, the test subject must have been interpreting the actual meaning of the calls rather than 'mindlessly' responding to their acoustic content. Both calls referred to the same social situation and habituation occurred simultaneously to both. By contrast, when the experiment was repeated using two calls that refer to very different contexts (e.g. the leopard and eagle alarm calls), habituation to one of the calls did not transfer to the other. These results indicate that the monkeys have some form of semantic, representational communication, which is a first step towards language, although human language involves much more than referential relations between words and objects or events. The point of interest here is whether they are aware of the meaning of the communication or merely act automatically in highly specific contexts. Unfortunately, these experiments cannot answer this question directly. The monkeys may be aware of the meaning but not necessarily.

Also, is the monkey that is making alarm calls aware of the state of knowledge of the other monkeys in its troop?

Cheney and Seyfarth say that it is not, because it will continue to give alarm calls long after everyone in the troop has seen the predator. I would suggest that this may not be a situation in which even humans would take cognisance of the mind state of others. In life-threatening situations most of us tend to focus in on our survival strategies; only some exceptional individuals act altruistically and show awareness of others. Under imminent attack from a predator the vervet monkey may focus attention and alarm call but not take into account the behaviour of the other troop members. This situation does not appear to be one on which to base general conclusions about the ability of monkeys to be aware of the state of knowledge of others.

Cheney and Seyfarth have tested macaque monkeys in the laboratory and reached the same conclusion that they did for the wild ones: that they are unaware of the state of knowledge of others. They investigated whether a mother responds differently when her offspring is ignorant of a situation compared to when it knows about it. One situation involved raising the alarm when a technician approached with a net, used to capture the monkeys, and the other involved calling to indicate the presence of food. The first situation mimicked the approach of a predator in the wild and, not unexpectedly, the mother gave the same type and number of calls irrespective of whether her offspring knew about the predator or not (different mothers and their offspring were tested). The same result was obtained for signalling about food. The mother called to signal the presence of food irrespective of whether the infant knew or did not know that food was there. It could be that testing the mother's behaviour when food is given is free from the problem of focussed attention under stress for survival, but the monkeys could have been so hungry that they were just as stressed in the test with food as they were in the test with the predator. Unfortunately, the researchers mentioned nothing about this and they did not measure any other behaviours that might indicate the level of stress. Cheney and Seyfarth have concluded that monkeys are unaware of the mind state of other monkeys even though the monkeys are astute observers of the behaviour of others and know the social relationships in the troop. I would interpret their results with more circumspection because the testing situations were or may have been arousing and stressful, the kind of situations in which even humans might not pay attention to the mind state of others.

Teaching

Teaching may be a manifestation of the ability to assess the mental state of another. It involves active participation in changing the behaviour of another. The teacher must recognise the difference between his or her own state of knowledge and that of the individual needing to be taught. There are reports of animals teaching another member of their species. Christophe Boesch has observed that mother chimpanzees in the wild sometimes teach their offspring how to crack open nuts. Chimpanzees crack the nuts by placing them on a rock or tree root, as an anvil, and then striking them with a hammer stone (discussed further in chapter 3). A mother performs this act more slowly when her offspring is looking. Boesch also observed a mother re-position her infant's nut on the anvil so that it could be cracked more easily. It appears that the mother not only taught the infant but also did so intentionally, acting with an understanding of the infant's specific lack of ability. Many primatologists use this example as evidence of mental-state attribution, meaning that the mother was able to attribute ignorance to her offspring. As David Premack says, the mother has a 'theory of mind'. This may well be so, but was the mother actually aware of the infant's mental ignorance or the infant's physical (skill) ignorance? The mother might have had no understanding of why the infant was behaving in a particular way and acted with the intention of changing the infant's behaviour, not its state of knowledge. This would be a less sophisticated form of attribution but it would be attribution nevertheless.

41

The chimpanzee Washoe, who learnt to communicate using American Sign Language, was given an infant chimpanzee to raise after her own baby died. She was observed, on several occasions, moulding the hands of the infant, Loulis, into signs. Washoe had been taught to sign by humans who sometimes moulded her hands and, apparently, she was using the same teaching method for Loulis.

Seyfarth and Cheney have reported that mother vervet monkeys do not appear to correct their young when they make inappropriate responses on hearing the various alarm calls: for example, standing up to look at the ground when they hear the alarm call 'eagle'. The mothers do not appear to encourage infants that have responded correctly to an alarm call and they do not appear to punish those that have responded incorrectly. The mothers do not appear to be aware of the mistakes of their offspring. Alternatively, they are aware of their infants' mistakes but they do not correct them.

Unfortunately, there is too little information on teaching in animals available to allow us to decide whether teaching in nonhuman species involves awareness of the mental state of another or whether simpler processes are being used. Some would argue that the absence of many examples of teaching indicates that it occurs only rarely in animals, as opposed to the common occurrence of teaching in humans, but I do not agree with this. Field ethologists tend to see what they are looking for and they overlook the behaviours that they have not thought about. This could be the case for teaching in animals because it has only quite recently become a topic of debate.

Reading another's mind state

The ability to know what another individual might be thinking or what another individual believes is an important aspect of awareness in humans. We can estimate and contemplate the state of mind of another individual. This ability is sometimes referred to as attribution of mental states to others or as having a theory of mind. As mentioned previously, there is evidence that children can attribute mental states to others by the time that they are two or three years old. How do our closest relatives, the apes, perform on tasks similar to those given to human children?

The primatologists Premack and Woodruff tested a chimpanzee on a task that might indicate this ability to read another's mind state. The chimpanzee was shown a series of videotaped scenes of a human actor struggling to solve a number of problems, such as reaching for a bunch of bananas or getting out of a locked cage. As well as seeing the videotape, the chimpanzee was given a series of photographs, one of which showed a solution to the problem. For example, a stick was included for the banana problem and a key for the cage problem. The chimpanzee chose the correct photograph to solve each problem, suggesting that she understood the actor's purpose, but she did this only when the actor in the videotape was her favourite trainer. When the actor was one that she did not like, she chose an incorrect photograph. It appears that she was intending to deceive the disliked trainer but, alternatively, it is possible that she only attended fully to the task when her favourite trainer appeared on the videotape.

More convincing evidence that chimpanzees can attribute mental states to others comes from the studies of Daniel Povinelli, at the New Iberia Research Center in Los Angeles, USA, and colleagues. Chimpanzees were required to attribute the mental states of 'knower' and 'guesser' to each of two humans. The chimpanzees were presented with four cups, one of which was baited with food. The knower was the person who had baited the cup in the presence of the chimpanzee being tested but without the chimpanzee being able to see which cup was actually baited. The guesser either waited outside the room while the cups were being baited or stood in the room with a bag over his head. At testing the knower pointed to the baited cup, whereas the guesser pointed to any cup at random. The chimpanzees were able to learn to act on the advice of the knower rather than the guesser, a result that the researchers interpreted as showing that chimpanzees are capable of modelling the visual perspectives of others.

These sorts of experiments provide convincing evidence that chimpanzees are aware of the state of mind of other individuals, and in these cases that they knew the state of mind of humans. It would now be interesting to see if other species can do likewise, although the manner in which they are tested would have to be adapted to meet the requirements of each particular species. Povinelli and colleagues have tested rhesus macaque monkeys in a testing situation very similar to that used for the chimpanzees and the results showed that they were unable to learn who was the 'knower' and who was the 'guesser'. Rather than being a failure of macaque monkeys, as compared with chimpanzees, to attribute mental states, this result could have been due to species differences in attention, or in social behaviour, or on the past experience of the particular animals tested. All of these factors that may influence performance on the task need to be considered before making any general statement about the ability of a species.

Deception

Social interactions are likely to be more complex in species that can empathise and 'read' each other's minds because awareness of the mental state of others would provide a powerful means by which to predict their behaviour. Social interactions would therefore be based on predictions or hypotheses, rather than being immediate responses to situations as they occur. The ability to assess the mental state of others and to predict their behaviour would also lay the basis for being able to deceive another intentionally; that is, to mislead another into believing something that is incorrect.

First let me give some anecdotal examples that *might* involve the use of cognitive processes for deception. Two monkeys were engaged in a fight; one moved away and

44

the other stretched out her hand as in a peace-making, contact gesture but, when the other monkey responded by taking hold of the outstretched hand, the first monkey grabbed hold of her and attacked again. Was the gesture made with the intention to deceive? An alternative explanation might be that, at the moment that the monkey put out her hand to make the gesture, she was motivated to signal reconciliation but, when the other monkey approached and made contact, she switched to aggression. Which is the more parsimonious explanation for the behaviour? A behaviourist would say the latter but, were we to substitute humans into this interaction, few would question that it was an act of deception. I want to point out that the interpretation of the behaviour that we will accept as being true depends on whether we believe that the species involved is capable of higher levels of cognition and consciousness. It is a matter of our attitudes to the species in question. On the other hand, the fact that this particular behavioural sequence is observed cannot, in itself, be used to prove the existence of higher cognition and consciousness in the species in question. There are many clever things that animals can do that do not require explanations based on higher cognition.

Let me give another example. I feed my three dogs together and one eats faster than the others. Having finished her bowl of food, on occasions, she will bark and run towards the gate as if someone were coming. The other two dogs follow and she dashes back to eat the food that they have left. It seems to me that she has played this trick too often to get away with it any more but it is, in fact, that repetition that makes me more convinced that it may be an intentional act of deception and not her own mistaken response to a sound at the gate or simply chance. There are two other aspects of the behaviour that lead me to deduce that it is deception using higher cognition: she would not hasten back to the food bowls before the other dogs if she had genuinely perceived that someone was at the gate, and she runs back to the food before she reaches the gate, leaving the other dogs charging to the gate alone. Not only has she managed to get the other dogs to leave their food but also she has contrived to make them fully occupied at the gate while she consumes the food that they have left. Of course, deception can occur only so long as the others are not aware of her false alarm and, if she repeats it too often, they will learn eventually. They will become aware of her intention to deceive (see chapter 3). This is likely to be why reports of behaviour that appear to involve deception are relatively rare. The difficulty is discovering acts that, although rare enough to deceive, are repeated enough not to be merely chance. It must be true that the more intelligent a species is, the fewer times a particular form of deception can be repeated without it being detected as a trick. It follows, therefore, that it might be harder to find convincing, repeatable evidence of deceptive behaviour in species that are more likely to use higher cognition to deceive.

Another, similar example of 'crying wolf' has been seen in the Arctic fox. An adult fox managed to steal a piece of food from a young one by issuing warning calls, on which signal the young fox dropped the morsel and ran off into the rocks. The adult then ate the food. This was repeated several times on different days.

Nor are such examples confined to mammals. Charles Munn has observed what he describes as deceptive behaviour in two species of flycatching birds (the bluish-slate antshrike and the white-winged shrike tanger) that he has studied for several years in the Amazon rainforest. These birds lead flocks of mixed species as they move through the rainforest canopy, acting as sentinels by giving alarm calls when bird-eating hawks are in their vicinity. In return, they feed on insects flushed out by the foraging of the rest of the flock. When an insect has been flushed out by a bird of the other species, the sentinel species joins in the chase to catch it. Munn has observed that, during the chase, these sentinels use the predator alarm call. He believes that they use it falsely to distract the other bird, even though only slightly, thereby gaining an advantage for capturing the prey. The 'false' calls were given in the absence of a hawk and during chases, not when on sentinel duty. It is possible, however, that the birds emit alarm calls when they are highly aroused, either on seeing a predator or during the food chase. That is, the alarm calls may be simply a read-out of the state of arousal and not intentional deception. Irrespective of the causation, the outcome for the flycatcher would be the same, an advantage in obtaining food. But one interpretation involves cognition, whereas the other does not.

As evidence against the interpretation that the bird is simply emitting the alarm call as an outcome of being highly aroused, Munn reports that the calls are not usually given when the birds are searching for prey alone. Unfortunately, this evidence is not conclusive because it is quite possible that the bird's state of arousal is higher during competitive chases than when it is foraging alone. By measuring heart rate or other physiological responses to stress the answer to this might be determined, but this would be very difficult to do in wild species, and it has not yet been done.

I have mentioned previously the rooster's use of the food call to attract a hen. Gyger and Marler have presented some evidence, although not comprehensive enough, that indicates that the rooster is more likely to use this tactic of deception when the hen is further away. According to these researchers, when food is actually present, the rooster is more likely to give a food call when a hen is nearby, reporting honestly to his audience that food is present. When food is absent and the rooster gives the food call to deceive, the hen is more likely to be further away. The reason for this might be that cheating will have a successful outcome only if the lie is not detected. If the hen were close by, she would be more likely to see that no food is present and therefore not approach. Moreover, cheaters might even be punished or, at least, ignored. We know from the experiments of Povinelli and colleagues, discussed previously, that chimpanzees can learn to ignore a cheater. I suspect that this ability occurs in many other species.

Sometimes animals remain silent in conditions in which they would usually emit calls. For example, many species of birds and mammals have been observed to emit food calls when a source of food has been discovered, and thus other members of the species gather in the same spot to feed. In some species, there are occasions when an individual does not call on finding food. Is this intentional deception, performed so that the food does not have to be shared, or has the animal failed to call for some other reason, such as not feeling particularly hungry or not preferring the type of food found? It is difficult to eliminate the alternative explanations for withholding information.

An often cited case of deception is the 'broken-wing display' of the ground-nesting plover. When a hawk flies overhead, the nesting plover runs away from her nest dragging one wing in a dramatic display feigning injury. This distracts the hawk's attention from the nest, as the predator is more likely to attack an injured bird. As soon as the predator swoops down, the plover flies away. Some argue that this is intentional deception, whereas others prefer to describe the behaviour as an unconscious response given to the signal 'hawk near nest'. There are more details to consider. To make the display the plover moves to a location close to where the predator is moving rather than where it was first sighted. While carrying out the broken-wing display, the plover also looks around to monitor the predator's behaviour and varies the pattern of the display to attract the predator. If the predator is not paying attention, the plover may approach and display more intently. Thus, the behaviour is not fixed or invariant, suggesting that the behaviour is not a totally automatic response triggered by the sight of a predator. Moreover, in an experiment using humans as potential predators, plovers learnt the individual characteristics of humans who had looked at the nest when approaching and they displayed more to them than to humans who had walked past without looking at the nest. The behaviour is by no means simple but it could be programmed by a set of rules. The plover's behaviour is definitely very clever but we cannot tell whether it involves higher cognition or is governed by a relatively simple set of rules. We might note, however, that the plover appears to be able to follow the eye gaze of the predator, because it displays more when a human predator looks at the nest. As discussed previously, in humans and other primates such eye-gaze following is considered to indicate self-awareness.

There are many anecdotal reports of deception in primates. In their field studies with baboons, Richard Byrne and Andrew Whiten have observed deceptive tactics used to obtain food from a dominant animal. A young baboon came across an adult about to eat a corm that he had dug from the ground, an activity that the young one may not have been able to do itself. The young one screamed loudly and its mother came running aggressively towards the adult with the corm. He dropped the corm and ran off with the mother in hot pursuit, and the young one proceeded to eat the corm. The researchers said that there was no doubt that the mother believed that her offspring had been hurt. This may be so, but it is difficult to know whether the young one actually used the scream deceptively. It might have screamed in frustration and the outcome may have been fortuitous. The researchers did say that the same individual was observed to use this tactic three times in several weeks, which might suggest intentionally but does not prove it.

There are many more examples in the scientific literature and more are sure to be added now that deception has become a much discussed topic. My opinion is that we do not yet have sufficient evidence that would prove that any of these acts are intentional deception based on cognition. Higher mental processes may, indeed, be necessary for some of the examples that I have discussed; the problem is where to draw the line. We are inclined to accept that deceptive acts performed by primates involve cognition and are intentional, but I would argue that the same may also be the case for some of the deceptive behaviour of birds and of other nonprimate mammals. In his book *The Thinking Ape* Richard Byrne claims that cognitive deception is largely confined to primates. He admits that domestic cats and dogs use deception frequently but he thinks that this results from their interaction with humans. Reports of deception in wild nonprimate mammals are rare. I would like to suggest that this apparent rarity may be simply a bias introduced by the main interests of researchers working in the field. Given that primates are closer to humans, field workers might be more inclined both to look for deceptive behaviours and to notice them when they do occur because they are more similar to the kind of tactics that we might use ourselves. In other words, the implied evolution of deception, and with it intelligence, in Byrne's claim may be misguided.

The difficulty in trying to use any of these reports of deception as evidence for cognition and awareness of others is that, although behavioural acts of deception must occur rarely to deceive, deception itself is not uncommon, even in lower species of animals. Many brightly coloured and patterned insects, for example butterflies and caterpillars, mimic the appearance of poisonous relatives so that they can ward off predators (birds) even though they are not themselves poisonous. This is deceptive mimicry but obviously it does not involve cognition. To make the distinction between this kind of deceptive mimicry and deception that uses social manipulation, the term *tactical deception* is used to refer to the latter. Again, where does one draw the line between one kind of deception and another?

Other species may use vocal mimicry deceptively to ward off predators or intruders encroaching upon their territory. Vocal mimicry uses brain mechanisms but maybe not cognition, since cognition requires higher processes that are not automatic. I do not wish to imply that vocal mimicry is not cognitive or that it is not intentional deception but I do wish to stress that, as yet, we do not know. The issue at stake may be not the ability to mimic but when and how to do it. Is it used creatively and differently in different contexts, or is it merely the sound equivalent of the visual mimicry in butterflies, given off automatically just as the brightly coloured butterflies ward off predators? At present, we have no answers. I will discuss this further in chapter 6.

Intentionality

Intentionality is planning ahead, anticipating the future. Intentionality is, unfortunately, another ambiguous term. Behaviour may appear to be intentional or have a previously planned purpose but the animal performing the behaviour need not be conscious of the planning or purpose. Many animals will go out in search of food at only those times of the day when it is available. Some species of bats, for example, wake up at dusk and go to catch insects, and they do this at a set time. At the time that they are awakening, they may have no thoughts of any plan to search for insects. They may simply wake according to an internal clock (referred to as a biological clock) and then go to feed automatically. If so, their behaviour may appear to be intentional, but no awareness or consciousness underlies it. They may simply be behaving like clockwork, as Descartes claimed. Of course, the bats may be conscious of their intentions but mere observation of their behaviour will not tell us that.

As I have just discussed, teaching and deception appear to the observer as intentional behaviours, but this observation alone cannot prove that they are conscious behaviours. There has to be a plan to change another's behaviour or to trick it purposefully. That is, we might say that the teacher or deceiver must have a 'vision' of the future.

Making a tool to be used for obtaining food may require planning ahead, but not necessarily. Chimpanzees and orang-utans are known to fashion tools for termite 'fishing' (tool using will be discussed further in chapter 3). When they are fashioning the tools, are these animals aware of the use to which they will put the tool? My intuition tells me that they are, but merely observing them engaged in this behaviour does not provide an answer to this question. Many species of rodents and birds store food in spring for future use in the winter. This seems like pre-eminent planning for a purpose but it may simply be unconscious behaviour triggered by a biological clock. If I had to guess, I would be inclined to say that most examples of food storage may not involve conscious intention, but that making a tool for a specific purpose may well be conscious. Unfortunately, there is no evidence that allows me to know which of my suppositions is correct.

Hunting by stalking prey may, perhaps, involve intentional planning ahead. It requires anticipation and planning to intercept the prey. Animals that can predict the behaviour of their prey more accurately will be more efficient hunters. The ability to mind-read another species is required to optimise hunting success. Depending on the species, this might be a more difficult task than reading the minds of members of one's own species. Awareness of other members of one's own species may be a direct extension of self-awareness. Awareness of the mental state of another species requires at the very least a translation of that ability to deal with the peculiarities of the other species. Of course, it may be possible to design a sophisticated machine that can hunt down certain species, but the intent observation of the prey and moment-to-moment adjustment of behaviour seen, for example, in lions hunting down zebra that they have singled out from the pack is complex behaviour that does not appear to be automatic. Perhaps it could be described by certain rules and perhaps the hunters follow these unconsciously, but I do not happen to believe that this is the case. This is my belief, others are entitled to theirs.

For species that hunt in packs (e.g. dogs and even chimpanzees), efficient hunting requires group co-operation and it may require mind-reading of the group members as well as of the prey. This is an extremely complex process. When chimpanzees set out to hunt down another primate to kill it for food, they appear to be doing so with intent. They use integrated strategies to corner their prey that cannot be completely preprogrammed. These strategies are certainly clever, if not conscious. The same appears to be the case in wild dogs, who stalk and kill their prey in groups. These highly social behaviours *appear* to be planned ahead (i.e. intentional) and we would definitely say that they were so were we observing the same behaviour in humans. To prove that it is the case in animals is far from simple. Again the problem of language intervenes: we can ask humans about their intentions but this channel of understanding mental processes is not available for animals. I have no hesitation in saying that group hunting looks like it involves conscious, intentional behaviour but, unfortunately, that does not prove that it does. However, it is not plausible to account for sophisticated and flexible behaviour in terms of stimulus–response relationships carried out on a moment-to-moment basis. Some of the actions of both humans and animals in these situations might occur as a result of rapid decisions without higher, conscious processes (e.g. according to simple rules, such as do B if A happens, and so on) but decisions about what, where and when to hunt and how to solicit and maintain group cohesion for the hunt are likely to involve higher mental processes and, probably, consciousness.

Suffering with others

I began this chapter by discussing feeling in animals and said that it is now commonly accepted that animals can feel physical pain inflicted upon them. Provided that it is within the capacity of a species for individuals to empathise with each other, a given individual may suffer by seeing another's suffering. Thus, provided that animals are aware of the physical, emotional and mind states of other animals, it is possible for one animal to suffer because it observes pain being inflicted on another animal. The suffering in this case would be emotional rather than physical pain.

Most scientists working in laboratories take little or no account of this possibility. I have often seen biochemists and neuroscientists killing rats while their cagemates watch. It may also be the case that individuals are aware of the emotional states of others. That is, one individual may suffer by being aware that another individual is suffering in ways other than physical pain. There will be more discussion of this in chapter 7.

There may be no single behaviour, yet known, that conclusively proves that at the least some animals have self-awareness or awareness of others, have intentionality or can attribute mental states to others, but overall we have indications that this is the case.

CHAPTER THREE

MENTAL IMAGES, MEMORY AND INTELLIGENCE

Some years ago I had a blind dog. She arrived from England at my house in Australia already blind and the first obstacle she had to negotiate was a flight of stairs leading up from the front door. She learnt to make her way up the stairs by running her snout across the width of each stair before stepping on it. This became a completely stylised or stereotyped behaviour. One day, however, she stood at the bottom of the stairs, not following as I called from the top, and she remained there motionless, as if calculating something. Then she suddenly took off up the stairs at a rapid pace with her head held high, without measuring each step that she took. From that time on she always used this new strategy to climb those stairs, although the measuring approach was used to negotiate other unfamiliar stairs. On that day when she changed the strategy she had gained insight into the problem. Insight is a form of problem solving that has been associated with higher intelligence, and it was once thought to be unique to humans. It is an aspect of intelligence and thus, in turn, it has been associated with awareness or consciousness.

A number of behaviours or cognitive abilities related to intelligence have been associated with awareness and consciousness. In addition to problem solving and insight, these are versatility, the ability to categorise objects and events, the ability to form concepts or rules and the ability to form mental representations of objects and events. Some of these abilities are related to each other and all of them rely on the animal's ability to form memories. I will discuss each in turn.

Intelligence or 'intelligences'

An animal with the ability for complex cognition is said to be intelligent. In chapter 1 it was mentioned that cognition and intelligence should not be confused. Cognition refers to those processes in the brain that use higher information processing. Although cognition and intelligence are linked, it might be better to reserve use of the term 'intelligent' to refer to the *behaviour* that is generated by higher cognitive processes, and thus distinguish it from the term cognition. In other words, complex cognition gives rise to intelligent behaviour. In solving a complex problem, for example, cognitive processes would be involved in finding the solution, and the behaviour that occurs as a result of solving the problem would be intelligent.

But, what do we really mean by intelligent behaviour? Not all behaviour that *appears* to be intelligent to the observer uses higher cognition. Animals, including humans, may exhibit such behaviour without it being a reflection of their intelligence. One might call this clever behaviour rather than intelligent behaviour.

Having made this distinction between intelligence and cognition, I must point out that the terms are not always used in this way. Many people use the term 'intelligent' to describe an individual rather than a particular behaviour. If an individual is 'intelligent', how does this show in his or her behaviour? At this stage we have reached a major controversy. Psychologists try to narrow down human intelligence by measuring the Intelligence Quotient (IQ) of individuals. There are a number of IQ tests, all of which are in the question and answer format. IQ, however, may have little bearing on problem solving or 'intelligence' in the world at large. There are, in fact, sufficient problems with measuring intelligence in humans to make us extremely wary of applying the term to animals.

In most publications about animals the term intelligence is used interchangeably with cognition. In fact, both terms are used in such a way that their meaning remains rather vague. I have to make it quite clear that, although the term intelligence is often used with reference to animals, and is so used in this book, there is no accepted, precise definition for it. Like consciousness it is a term that cannot be defined in a unitary way. It would be pointless to come up with some battery of tests that might attempt to measure in animals the equivalent of IQ in humans because animal species vary so much in their senses, their manner of processing information, and so on. We do, however, recognise that an animal with a greater cognitive capacity is more likely to display intelligent behaviour and more likely to have consciousness than one with a smaller cognitive capacity.

When referring to humans, usually we apply the single term 'intelligence' to a diverse set of activities that we assume are controlled by a common set of cognitive processes. There is, in fact, no evidence that this is the case. Furthermore, there is no evidence that different species use the same cognitive processes to carry out similar types of behaviour.

As a general rule, we consider animals that are more like us as being more intelligent, but it is important to recognise that each species is adapted to its particular environmental niche and performs 'intelligently' in that niche. If we think of intelligence in this way, it is pointless to classify one species as more intelligent than another. This seems a reasonable position to take. One could say that there are many different 'intelligences', rather than ranking all species on the same scale of intelligence. Some species that may appear to be less intelligent than others when they are all tested on the same, rather arbitrarily chosen task (e.g. going around a barrier to reach something on the other side) may perform very 'intelligently' on tasks better suited to their own specialised abilities. It would be better to see intelligence in terms of the entire repertoire of the behaviour of a species and in the ability of the species to establish new relationships and to solve novel situations but, unfortunately, we have little information about the breadth of the potential behavioural repertoire of many species.

To move from intelligence to consciousness, it is assumed that consciousness comes about only when a certain level of intelligence is reached, that is, when a certain level of cognitive complexity is reached. Not all species can be conscious, or conscious in the same way, even though every one may be perfectly adapted to perform intelligently in its own niche. The issue then is when and in what species did cognitive complexity or intelligence reach a level at which consciousness could emerge?

The matter is complex because, by and large, increasing complexity is seen as following a linear or hierarchical path. As animals evolved their brains and their behaviour may have become more complex, but evolution has not occurred exactly in a linear fashion. The evolutionary tree has branches at which one line branched from another. For example, reptiles evolved from amphibians and both birds and mammals from reptiles. We see the mammalian line of evolution as the trunk of the tree, because eventually it led to humans, and birds as being on a side branch of the trunk. Birds went along their own separate path of evolution and, as we shall see later, they developed cognitive complexity and intelligence of a kind different from that of mammals. Instead of seeing the branches of the tree of evolution as lesser than the trunk, these days some of us prefer to refer to an evolutionary vine, rather than a tree, in order to recognise the differences between species but not to place them in a hierarchy. Different 'intelligences' have arisen on different branches of the vine, many times over.

Has consciousness arisen once only or more than once on different branches of the evolutionary vine? Birds, for example, with their different complex cognitive capacities may have evolved consciousness quite independently of mammals. If so, would their consciousness be the same as that of mammals or quite different? Like intelligence, consciousness might differ according to the species and its environmental niche. As with intelligence, we might overlook those forms of consciousness that are too different from our own.

I have raised these points only to show that this 'thing' we call consciousness, like the 'thing' we call intelligence, is unlikely to be unitary or fixed. There may be certain environments that are more likely than others to bring out intelligence and consciousness of a certain kind. According to Alison Jolly of Rockefeller University, USA, and Nicholas Humphrey of Cambridge University, USA, the greater intelligence of higher primates evolved to deal with the problems of social life. It would be only in social life (be this social life within the same species or between species) that deception could occur and the ability to predict the behaviour of others would be particularly beneficial (see chapter 2). Thus, social intelligence, and consciousness, might be used for social manipulation. Humphrey argues that social intelligence is used also for shared knowledge of the habitat and of techniques used for finding food, building nests, and so on, and for transmission of learnt information (culture). He says that, with increasing time spent on social activities, the members of a species have less time to spend on other subsistence, nonsocial behaviours. They must therefore become more efficient in performing these latter activities, and this adds to the intellectual demand. With social and nonsocial demands for increased intelligence, a snowballing effect occurs and the evolution of intelligence gets extra impetus. Although interesting, this hypothesis is not watertight.

Social complexity might well provide a powerful demand for intelligence and, eventually, consciousness but, based on the research that my colleague Gisela Kaplan and I have done on orang-utans, I do not think this is a complete explanation. Orang-utans are solitary apes compared with chimpanzees and gorillas but they are not less intelligent. There is a saying that, if you give a screwdriver to a captive chimpanzee, it will throw it out of the cage; give it to a gorilla and it will scratch itself; give it to an orang-utan and it will use it to unscrew the cage and escape. Certainly, in tasks requiring any form of manipulation orang-utans excel. This might be merely anecdotal evidence but even Humphrey has remarked that orang-utans do not fit his hypothesis. Higher intelligence might be demanded by environments that require much decision making and learning of the skills for survival. It has been suggested that wild orang-utans use a large amount of their cognitive capacity to negotiate their way through the canopy. With such heavy bodies they must be constantly assessing which boughs can support their weight, and an accurate decision on this matter would depend on much learning about the strength and subtleness of boughs. Thus, life style, social or otherwise, may demand intelligence and perhaps consciousness too.

Versatility/adaptability

Versatility is an aspect of intelligence. Biologists tend to use the term adaptability to mean the same thing as versatility. Some species are specialists, able to live in a narrow range of conditions and eat a narrow range of food, whereas others are more adaptable, being capable of adapting to many different conditions and food types. Humans are highly adaptable as we have spread to a multitude of different environments in all parts of the world, but so too have many insects, such as cockroaches. Adaptability does not necessarily have anything to do with intelligence, but intelligence may assist some forms of behavioural adaptability. Humans have managed to inhabit inhospitable regions of the earth by using their intelligence to construct shelters, make clothes, obtain food, and so on. Here our mental abilities have permitted versatility or adaptation.

Adaptability is a concept that is only tenuously related to intelligence, but it is a term that has come into greater use in relation to artificial intelligence as well as the intelligence of animals. Hence the need to discuss it here. Adaptability may be a characteristic applied to an individual or the individuals within a species. Gisela Kaplan (of the University of New England) and I have been inclined to say that the intelligence of orang-utans is manifested in their ability to adapt to different environments. As discussed in chapter 2, orang-utans in rehabilitation centres adapt to interactions with humans by using their tools and imitating their behaviour. This is well known to labourers working in rehabilitation centres for orang-utans: as mentioned previously the orang-utans may 'help' by taking the shovel to dig the garden, and the paint-brush to paint the walls, the floor and perhaps the roof, and they take the saw to attempt to imitate sawing wood.

Adaptability applies to individuals that can solve complex problems and may be able to plan ahead. Adaptability is also applied to the evolution of a species as it adapts to a changing environment. Some scientists, such as Jonathan Schull of the department of Psychology at Haverford College, USA, say that this means that species are 'intelligent'. He suggests that biological species and intelligent animals have much in common in their abilities to adapt to their respective environments and in how they interact with other species or individuals, respectively. In this sense all species from ants to apes are 'intelligent' as long as they are adapted to their environment. This very broad use of the term 'intelligence' is entirely separate from intelligence generated by higher cognitive processes. It is, therefore, not useful in our discussions of intelligence related to complex cognition and consciousness, but it is important to keep it in mind.

Problem solving and insight

The ability to solve problems is considered to be an aspect of intelligence in both humans and animals. There are many ways to solve problems. The simplest one is by trial and error, in which every possible strategy is tried at random and the solution to the problem is found by chance. This approach does not necessarily require higher cognitive processes, although they may be involved. The most sophisticated way of solving a problem is to use insight. In this case the subject thinks about the problem and uses prior knowledge of a different situation to come to a solution without trying out any other ways of dealing with the problem. When we have such an insight, we say that the solution 'came in a flash' and we feel a sense of pleasure (sometimes referred to as an 'ah ha' feeling).

Some people think that insight is one of the important characteristics that separates humans from other animals. It is difficult to design experiments that would prove beyond doubt that an animal is, or is not, capable of insight but I believe that many researchers rather too hastily assume that problem solving by animals is imitation rather than insight. It is true that there are very few reported examples that might indicate insight in animals, but we should remember that insight is considered to be an aspect of learning and the field of learning in animals has been dominated by 'learning theory', in which experimental psychologists study the kind of learning that results when a particular response is rewarded (e.g. by giving a food reward) or punished (e.g. by applying an electric shock). For example, a rat can be trained to press a bar when a light comes on by rewarding it with a pellet of food each time it presses the bar. At first, it presses the bar simply because it is something to do and it does not know that it is associated with food but, after many trials (of pressing the bar and being rewarded with food), it will learn to associate bar pressing with food. This is called conditioned learning. The same sort of training procedures are used frequently by circus trainers: in this case the animal is rewarded for performing a particular antic. Other sorts of learning that do not require any obvious reward or punishment have been largely ignored by experimental psychologists. Another example of learning with no obvious reward is imprinting, a powerful form of learning by young chicks and ducks, as well as by other species that are born in a relatively advanced state of development. By the process of imprinting they learn to recognise their mother and so follow her. Imprinting learning has been largely ignored by experimental psychologists but not by ethologists, who recognise it as a special form of learning essential for survival of the species. Insight learning, like imprinting, is carried out without food reward or punishment and it requires contemplation that may not be encouraged by most laboratory testing situations.

There are some reported examples of insight learning in apes. Lethmate describes the following sequence suggesting insight in a young orang-utan. The orang-utan was given a long rod which could be inserted into a transparent plastic tube to reach a sweet and push it out. The orang-utan knew what the sweet was but he did not know how to use the rod as a tool to obtain it. At first he bit the tube and tried unsuccessfully to insert the tool. He then moved away and sat down, apparently in frustration as he began to perform stereotyped (repetitious) behaviours with the tool and blanket. Then he glanced back at the tube and, apparently, at this moment the insight came to him. He got up, walked over to the tube carrying the rod, inserted it into the tube and obtained the sweet. Although he was, of course, rewarded by eating the sweet, this was only at the end of the sequence and his solution to the problem did not have to be conditioned by giving him lots of rewards during the learning of the task. Instead, the problem appeared to be solved in a flash of insight.

Experience of playing with objects may provide the basis for insight. A chimpanzee that has played with boxes of various sizes is more likely to show insight in stacking the boxes, smaller ones on top of the larger ones, to make a tower to climb up so that it can reach a bunch of bananas hanging from the roof of its cage.

Another possible example of insight learning may have initiated the washing of sweet potatoes in the sea by

Japanese macaque monkeys, which they do before they eat them. On the island of Koshima the macaques are fed by people who dump sweet potatoes, wheat and other food stuffs on the sand. Many years ago the scientists working with these macaques noticed that one of them was taking her potatoes to the water and washing the sand off before she ate them. In time, other members of the troop adopted the same behaviour, either because they imitated the first monkey or because they discovered the behaviour independently. Here we are interested in the first monkey's discovery of washing potatoes. If she came across it by chance, simply because she happened to go into the water when she had a potato in her hand and then dropped it, the acquisition of this new behaviour would not reflect any remarkable ability to solve the problem of removing the sand from the food. If, however, she knew that water could be used to wash things or parts of her body and then she applied this knowledge to the potato problem, she would have used insight. Without detailed observation of the initial performance of this interesting behaviour, I am afraid we cannot decide which of these explanations is more likely. But we do know that later the same monkey began to wash wheat in the water and that this practice also spread through the troop. This second discovery might suggest that this particular monkey has superior insight ability, because it is unlikely that the same monkey would have learnt twice by chance, unless she has some other peculiarity of behaviour which, say, takes her to the water more often than the other monkeys in the troop. However, a third of the troop of monkeys were also going to the water to wash their potatoes by the time wheat washing was discovered by only one monkey, and that was the same one that had discovered potato washing.

In an attempt to observe the processes of learning that may lead primates to wash their food, Elisabetta Visalberghi of the Instituto di Psicologia in Rome, Italy, and Dorothy Fragaszy of the University of Georgia, USA, gave sand-covered food to groups of capuchins (South American monkeys) and crab-eating macaques. These monkeys had water in their enclosures and the macaques were used to standing and playing in it. The capuchins were more hesitant about the water at first but later they played in it. Most of the macaques soon learnt to wash their sandy food before eating it but it appeared that they learnt to do so rather by accident as they took food with them when they ran to the water to play. The capuchins behaved differently. At first they sampled the sandy food and, finding it distasteful, tried to rub off the sand. Very soon (within the first six minutes) one of the capuchins began to wash the food in water before eating it and the researchers said that he appeared to do this 'deliberately'. He would take a piece of sandy fruit from a basin, go to the water to wash it, eat the fruit and then repeat the sequence. He also inspected each piece of fruit after it had been dunked in the water and washed it again if all of the sand had not been removed. It is rather unlikely for this behaviour to have appeared purely by chance. Insight learning for a deliberate purpose seems more likely. The other four capuchins in the same group acquired the behaviour later on and thus it is unlikely that they did so by insight. They may have imitated the first capuchin's behaviour, but a repeat experiment on a larger group of capuchins found that only some of the subjects learnt to wash their food. Depending on the social group and past experience of the animals, food washing may spread at different rates through the group. Regardless of this, in both groups of capuchins there was one individual or a few individuals who rather rapidly showed the behaviour of washing the food, and these few may have acquired the behaviour by use of insight with a plan in mind. The crab-eating macaques, on the other hand, may have acquired the same behaviour by the chance association of food and water in play. But, given that laboratory living and other social factors in the group may influence the behaviour, I would be reluctant to say that these differences are characteristic of the species and I would be equally reluctant to apply these results to the potato washing of the wild Japanese macaques. Nevertheless, these observations do point out the variety of ways in which monkeys can learn and the complexities involved in interpreting exactly what processes are going on.

Now I would like to consider some other forms of learning that show how clever animals can be. I have mentioned how rats can be conditioned to press a bar for a food reward. Using a similar procedure, pigeons can be trained to peck a key for a food reward. Pigeons can also be trained to peck at a key with a particular colour and avoid one of another colour (e.g. peck a red key for a food reward and avoid a green key, because pecks at green are either not rewarded or are punished), and they can be trained to peck at a key that has a particular pattern displayed on it and avoid one with another pattern. They can also be trained with three keys, each with a pattern displayed on it. The centre key provides no reward or punishment if it is pecked, and on it is a pattern that is matched by a pattern on one of the side keys. The key on the other side has a different pattern on it. The pigeon has to learn to peck the side key with the matching pattern to get a food reward. The side key on which the matching pattern occurs is changed randomly between the left and right sides on each pecking trial so that they pigeon does not learn simply to peck the key on, say, the left rather than the pattern. This is known as a matching-to-sample task. Once trained in this way, the pigeon can be tested for its ability to solve a variety of problems. It turns out that pigeons are remarkably good at solving very complex problems using these visual displays on the keys.

Using this method, Juan Delius of the University of Bochum, Germany, has shown that pigeons have an astounding ability to perform mental rotation problems of the type included in intelligence tests for humans. The pigeons were first trained to match-to-sample an abstract shape presented on the central key (Fig. 3.1). One of the test patterns was identical to the sample and the other was its mirror-image. Pecks at the matching stimulus were rewarded with food, whereas pecks at the mirror-image were

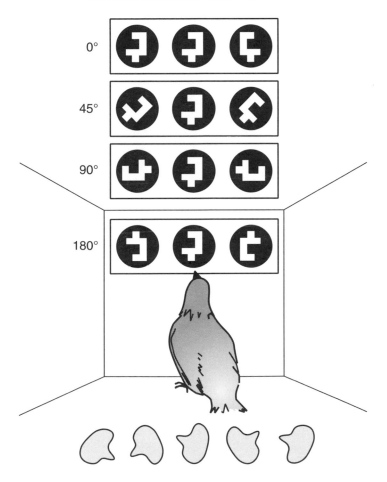

Fig. 3.1 A pigeon has an excellent ability to recognise symbols rotated at different angles. The pigeon has to peck the key (left or right) that matches the pattern displayed on the central key. The problem is similar to the standard rotation problem (at the bottom of the figure) of an intelligence test for humans *Source:* Adapted from Delius, 1987.

punished by a brief period of darkness. In training, several different shapes were presented all at the same angle of orientation. In testing, the pigeons were presented with shapes rotated at various angles relative to the sample. They were able to perform the task just as accurately and as rapidly as before. In fact, there was no decline in their ability to perform the task when the patterns were rotated. Humans tested on the same task (touching rather than pecking the keys) showed a significant decline in accuracy when the patterns were rotated and they also took longer to make a decision about which key to touch. Delius said that the pigeons were geniuses in comparison with the humans! Of course, this may mean that pigeons solve the problem using quite a different cognitive strategy, possibly related to their experience of looking down on objects in a horizontal plane and thus with no preferred angle of orientation, but their strategy is clearly not an inferior one.

Categorisation and concept formation

Pigeons further illustrate their highly developed cognitive capacities by being able to form perceptual concepts, such as those required to recognise different forms of trees, leaves, persons, water or fish in different contexts. Delius trained pigeons to peck at any key that had water on it regardless of whether the water was a droplet on a leaf, a lake, a glass of water, and so on. They were able to perform this task, according to Delius, by forming an abstract concept of 'water' recognisable in all of these different forms and contexts. They could do the same for trees of different kinds, as well as people and so on.

Pigeons can even use the abstract concept of 'sphericity', as determined by conditioning them for pecking at solid, three-dimensional objects, such as pebbles, bolts, pearls and buttons, instead of pecking at keys. The three-dimensional objects were presented on a series of metal plates attached to an automated system that moved them through the cage as the pigeon pecked. Each pigeon was presented at any one time with three objects on keys, either two spherical objects and one nonspherical or one spherical and two nonspherical. It received a food reward for pecking spherical objects and no reward was given when it pecked nonspherical ones. Presented with eighteen objects of each type, the pigeons learnt to perform the task within remarkably few trials. They were then tested to see whether they had acquired the concept of 'sphericity' by presenting them with over one hundred novel spherical and nonspherical objects. They were able to generalise to the novel objects, recognising them according to the abstract characteristic of 'sphericity', just as do humans, and they could even judge sphericity in photographs of the objects.

Pigeons can also acquire a perceptual concept of symmetry, an ability that is said to underlie the expression of art by humans. Delius showed that they can learn to discriminate between symmetrical and asymmetrical patterns and, once they have learnt this, they can apply the concept of symmetry to other types of stimuli that they have not seen before. They form an abstract concept of 'symmetry'.

There is also evidence that pigeons are able to solve problems by using abstract rules, such as 'oddity' or difference in terms of the shape of stimuli. They can learn to detect the odd stimulus in a group and generalise the abstract rule learnt to other types of stimuli. The same ability to perform oddity learning has been shown in primates, dolphins and members of the crow family.

Categorisation and concept formation have been shown in a very special parrot, named Alex. Alex has been trained by Irene Pepperberg of the University of Arizona, USA, to use English words to name objects and feelings. He can use a vocabulary like that of the sign-language-trained chimpanzees and he can identify, request or refuse more than one hundred objects of various colours, shapes and textures. For example, the experimenter may show Alex a green wooden block and ask 'What colour?' and 'What shape?', and he can answer each question correctly. He also expresses desires (such as 'I want peanut' or 'Come here').

Alex's ability to categorise or see the relationship between objects can be tested by presenting him with different objects and asking him to say whether they are the 'same' or 'different'. For example, he might be shown a blue wooden square and a blue paper square and, when asked 'What's same?', he replies 'Blue' and, when asked 'What's different?', he replies 'Shape'. Chimpanzees have been tested on similar tasks and Alex performs as well as they do. The concept of same/different is an abstract one, as arbitrary symbols must be constructed to represent the relationships between objects. Therefore, it relies on higher cognitive processes, and we can say that Alex exhibits intelligent behaviour. His behaviour is almost certainly more than merely clever, and this is a convincing way to demonstrate it in the laboratory.

To survive in the wild, animals must rely on well-developed capacities to categorise items, be that foods versus nonfood or familiar songs of other birds versus unfamiliar ones and the ability to recognise same versus different would also be important in social communication using vocalisations. Animals must also be able to recognise quantity. It must be within the capabilities of most species to recognise more versus less (e.g. more food versus less), but we know that at least some species can count. Alex can count up to six. When asked how many objects there are on a tray, he can say the number with an accuracy of about 80 per cent. He has a concept of numbers.

It is interesting that Pepperberg has reported Alex's performance with up to only six objects because seven seems to be a 'magical number' for animals as well as humans. Jacky Emmerton of Purdue University, USA, and Juan Delius, whom I have mentioned already, tested the ability of pigeons to discriminate 'more' versus 'less' dots presented on the keys of a conditioning box. They could distinguish one dot from two with 80 per cent accuracy and two from three and so on up to seven from eight, with decreasing accuracy as the numbers increased. In fact, at seven versus eight their accuracy had dropped to chance levels. They could not make this discrimination. The same drop in performance has been found in other species and even in humans tested on exactly the same task as the pigeons. Although the pigeons could be counting the number of dots on each key and then comparing them, Emmerton and Delius think that this is unlikely in this sort of experiment. Rather, the pigeon may look at the array of dots on one key and remember that briefly while it compares it with the array of dots on the other key. That is, they may form internal representations of the visual images on the keys. Whatever strategy is being used, the pigeon can make abstract discriminations based on numerical quantities. Primates can do likewise and Sarah Boysen of Ohio State University, USA, has demonstrated that a chimpanzee called Sheba can carry out some algebraic calculations, such as simple addition, using the Arabic symbols of numbers which we use.

Memory abilities

Pigeons must have an extensive memory to perform the tasks already mentioned and on some tasks their memories rival those of humans. Von Fersen and Güntürkün trained pigeons to remember hundreds of different patterns projected onto the keys of a conditioning box. The pigeons were rewarded with food for pecking one hundred different patterns, and they had to discriminate them from over six hundred other patterns that provided no reward when pecked. This discrimination is extremely difficult for humans, but the pigeons could learn to do it with great accuracy and retained the memory for it with an 88 per cent accuracy after seven months. This is remarkable.

Pigeons can also remember that they have seen up to 320 slides of (human) holiday scenes after a delay period of two years. Delius believes that they may achieve this astounding feat of memory by coding or labelling the information, possibly in much the same way that humans do so by using descriptive words. Other researchers, however, claim that the pigeons must use rather simple mechanisms to make these enormously complex visual classifications. Further experimentation will be necessary to find out the answer, but neither explanation detracts from the impressive memory and discrimination abilities of the pigeon.

Birds that store their food (parid and corvid species) also display remarkable memories. John Krebs of Oxford University, UK, has shown that European marsh-tits can retrieve their stored caches accurately at a large number of sites days after they have stored them. Some species in very cold climates even remember where their many caches are located from autumn until the following spring—and they store several hundreds of seeds over a period of just a few weeks.

I have deliberately chosen examples of memory capacity in birds because, until quite recently, this aspect of birds has been rather ignored. There is considerable evidence that other species form many, complex memories that persist over time. The much stated adage that 'Elephants never forget' is consistent with experimental findings, but elephants are not likely to be alone in having this characteristic. Many readers will be familiar with the fact that their pet dog or parrot may take a like or dislike to one of their friends and remember that particular person even after very long periods of absence.

For most species, having a long memory is a matter of survival. Orang-utans, for example, remember where their favourite fruiting trees are located and when the fruit ripens, as they return to particular trees at just the right time at each fruiting season. Such behaviour is typical of many species. Others can find their way year in and year out over enormous distances, following remembered paths. These are specialised skills that certainly rely on cognition and detailed memories. In fact, the need to forage for food is considered to be a driving force for increasing the cognitive complexity (or cognitive capacity) of the brain. On this basis, some people argue that ungulates (horses, cows, sheep, and so on) have had no pressure to evolve higher cognitive powers because they do not have to go out in search of food in the same way that species with more specialised diets must. The implication is that ungulates are less intelligent than many other mammals, but I would suggest that such beliefs are based on inadequate understanding of the cognitive abilities of ungulates. Furthermore, ungulates do not simply eat every blade of grass that they come across. They select favourite grasses and may even go in search of them.

It has also been hypothesised that the apes that stayed in the trees eating fruit experienced no evolutionary pressure to evolve higher cognition and that it was the descent of our ancestors from the trees and their shift in diet involving hunting for food that led to the evolution of hominids (the line of evolution to modern humans). I will discuss this more in chapter 5.

Other memory abilities must be applied to social situations. In chapter 2, I mentioned imprinting in young chicks. The chick learns the features of the hen and also of its siblings, and it remembers these for a very long time, possibly for the rest of its life. At first it forms a memory of the hen and follows her when she moves away from the nest. It also learns to recognise its siblings and can tell them apart from other chicks. Later it becomes sexually imprinted on the hen and this determines its preference for a mate in later life. It is these stable and powerful memories that direct its social behaviour. Chickens, when young and adult, must remember their positions in the social hierarchy (the pecking order) and to do this they must recognise other members of their social group so that they can behave appropriately when they encounter them. None of these memories are simple. For example, the hen must be recognised by her main visual features as well as her vocalisations and the way she moves. Her smell may be important also, as it is known that chicks imprint on certain odours. The hen must be recognised in different environments (that is, she must be recognised against a changing background of visual images, sounds and smells). These

73

memories are recorded in the chick's brain and they must be, as it were, written down according to some sort of chronological sequence that becomes a unique autobiography of each individual chick.

Similar memories are used by all animals as a basis for their social behaviour. As I have mentioned earlier in this chapter, some primatologists believe that social behaviour provided the evolutionary pressure to increase cognitive sophistication and, eventually, led to self-awareness. Although this hypothesis may have some validity, it should not be limited to the primates. All too frequently primatologists and some psychologists ignore the fact that many other species of animals have complex social organisations equivalent to those of primates. It can be said that, for all mammalian and avian species, the larger a social group is, the more complex the memories that each individual must hold and the more often those memories have to be updated.

Overall, the memory abilities of animals do not differ from those of humans. The memories of animals can be detailed and extremely stable. They can also be updated and they are essential for survival. It is possible that species, and individuals too, have memories that vary in their richness and that this is directly related to their cognitive capacity, but we have yet to discover this. Although the ability to form memories is a measure of cleverness or intelligence, it does not necessarily prove the existence of consciousness. Memories may be used to direct behaviour without the animal being conscious of them, just as a computer stores memories that direct the way it functions.

We can recall our memories when we wish, outside of any direct context related to the particular memory. They come into our consciousness and we can contemplate them. Can animals do the same thing? According to Merlin Donald of Queen's University, Canada, they cannot. He believes that even apes are unable to recall memories independently of triggers in the immediate environment. That is, Donald believes that they cannot recall memories at their will and carry out independent thought. I consider this to be a particularly prejudiced position to take, given our inability to access what an animal is thinking through use of language. In fact, Koko, a gorilla taught to communicate using sign language, does communicate how she felt in past situations (e.g. she expresses sadness when asked to recall her feeling about a lost companion, as will be discussed further in chapters 6 and 7). Of course, it could be said that this response was triggered by being asked the question, but we do not have access to times when she might have similar recall of her feelings without being prompted. Does she perhaps express her private thoughts in sign language? Even if she does not, that would not prove that she does not have private thoughts because, after all, we would rarely speak aloud our private thoughts. In the absence of evidence, people like Donald, who categorically state that all animals are locked into thinking about and responding to only the immediate environment, are expressing their attitudes to animals, not scientific evidence.

Mental representations

The human mind forms internal representations of objects and events. These representations take on a presence in the mind. We use them as a basis for communication by language and to make symbolic art forms, also used in communication. A sculpture or a painting may be the physical manifestation of the artist's internal representation. This does not mean that there is an exact picture in the mind. Mental images are elusive, invisible and have no objective existence like television images, paintings, photographs. Mental images do rely on certain physical processes in the brain, the activity of neurons, but they cannot be explained directly by the known physical processes of the brain. We also form mental images of sounds, smells and the feel of objects, and so on. They are part of memory, imagination and dreams and they may also be hallucinations. Even though we are able to describe visual images that we have in mind and have a sense of actually seeing them 'in the mind's eye', they are subjective and cannot be pinned down into any physical form.

Mental representations are an aspect of consciousness and they may be the basis on which symbolism and art developed. The ethologist Irenäus Eibel-Eibesfeldt, at the Max Planck Institute in Germany, considers that certain aspects of the perception of art as aesthetic are based on sensory processes that have a long evolution, and are therefore shared by many species of animals, but the creation of art, he believes, is unique to humans. I am not sure that we need to be categorical about this. What is art and what is not is dependent on the observer and that observer's ability to read the symbols. The topic of symbol use by humans will be discussed further in chapter 5.

The ability to form and use mental representations must require a highly developed cognitive ability, but the question of when the ability evolved remains open. There is evidence that it evolved much earlier than anthropologists seem to accept. Of course, humans may be unique in the way that they use mental images in communication, but it is unlikely that we are alone in our ability to form representations of objects.

Mental images of hidden objects

When we are searching for something that we have lost, we are able to 'visualise' the object in the mind. The mental representation of the lost object becomes paramount in our minds so that we may overlook other objects that we encounter during our search. We are said to have formed a *searching image*. Human infants of less than eight months of age will not search for objects hidden from them. The famous psychologist Piaget said that they have not yet developed 'object constancy'.

Object constancy is said to indicate the ability to form a mental representation and, surprisingly, even young chicks appear to be able to do this. Giorgio Vallortigara of the

University of Udine in Italy has tested young chicks on tasks in which they have to go around a barrier in order to get close to an object on which they have imprinted. Each chick was raised with a red table-tennis ball hanging in the cage so that it imprinted on that instead of on the mother hen. Once imprinted, a chick will always approach or follow the imprinting object so that it remains close to it. It treats the object as if it were a social partner. The chick becomes distressed when it is unable to be near the imprinting object. Thus, a chick imprinted on a red ball would follow after the ball and go around barriers to get to it. Vallortigara tested the chick's ability to form a mental representation of the red ball by putting the chick inside a small cage with transparent walls and placed inside a large circular arena (Fig. 3.2). From its cage the chick could see two screens placed at equal distances from its position, and the red ball on which it had imprinted. While the chick watched, the red ball was moved behind either one of the screens. The chick was held in the cage for two or three minutes longer and then released into the arena. If it could not remember which screen the ball had disappeared behind, that is, if it had been unable to form and store a mental representation of the object going behind or being behind, the chick would have approached either screen at random. It did not. All of the chicks tested approached the screen behind which the ball had been hidden from their view, and went around it to make contact with the ball. In another test, the same researcher found that chicks would walk around a short maze of corridors in the correct direction to be able to see the red ball through a small window. As each chick was making its way around the corridors it must have been orienting itself by using a spatial representation of where it would find the ball. In other words, the chick was aware of the existence of the ball even though it was not visible to the chick while it was walking through the maze.

In these experiments, the chicks were able to retain the mental representation for only two to three minutes. With

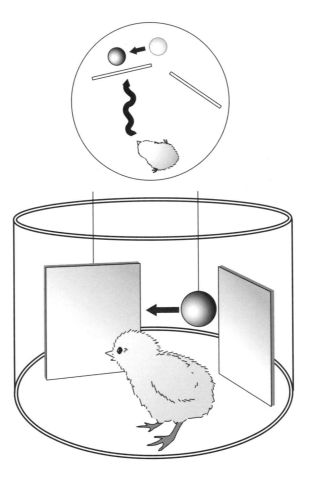

Fig. 3.2 A young chick has been raised with a red ball hanging in its cage and becomes imprinted on it. Here the chick is tested to see whether it can remember which opaque screen hides the ball. The chick is allowed to watch as the ball is moved behind one of the screens and a little later it is released into the arena. The chick approaches the screen hiding the ball and goes around it to find the ball. (Drawing not to scale.) *Source:* Experiment by Regolin and Vallortigara, 1995.

longer delay periods between seeing the ball move behind a screen and being released into the arena, they approached either screen at random. Therefore, while chicks can form mental representations, perhaps they are unable to retain them for long periods. This may be a consequence of their young age (adult fowls have not been tested for this ability) or because the species lacks the ability to make long-term representations.

Mental representations are also used to recognise visual objects when only a part of the object can be seen. Most objects in the world are opaque and thus we cannot see all of an object at once. The front hides the back, and other objects get in front of the one that we might want to see, and so on. Humans have no problem with this: we do not perceive only the separate fragments of the object but recognise the whole object when we can see only parts of it. We generate a mental representation of the nonvisible parts of the object. This ability would seem to be critical for all living species because prey as well as other members of the species are often only partly visible, being obscured by bushes or other barriers. Lucia Regolin of the University of Padua, Italy, and Vallortigara have shown recently that young chicks that have been imprinted on a red cardboard triangle (a two-dimensional coloured triangle cut-out placed in the cage) can recognise this triangle when it has a black bar through the middle of it (Fig. 3.3). They treat it as a partly obscured triangle and will approach it in preference to a triangle with no middle section in the region that would have been obscured by the bar (i.e. fragments of the triangle that would be actually visible on either side of the bar). By showing different combinations of the triangle and the bar, Regolin and Vallortigara have been able to demonstrate that the chicks are able to recognise the triangle when it appears to be partly hidden behind another object, the bar. The chicks could complete the mental image of the triangle when it was partly occluded. The chick, it would seem, possesses abilities to recognise partially occluded objects very similar to the abilities of humans. In fact, it might be

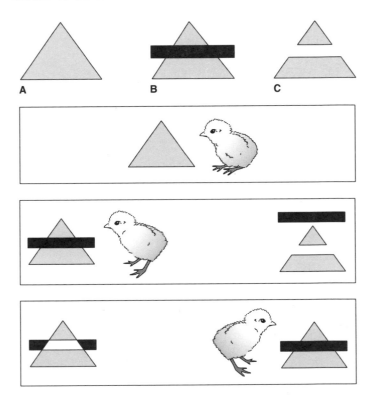

Fig. 3.3 A chick is raised in the presence of a triangular shape (A) on which it imprints. When tested with a choice between the triangle partly hidden by a black bar (B) and a triangle with the region covered by the bar missing (C) and also with the bar over the top or on each side of the triangle, the chick approaches the partly hidden triangle (B). This result shows that the chick is able to recognise an object when it sees only part of it *Source:* Adapted from Regolin and Vallortigara, 1995.

said that the visual capabilities of birds rival those of primates. However, mice can also complete mental images in the same way. Therefore, although this was once thought to be an ability unique to humans, it appears to be widespread amongst animal species and to have evolved very early.

Newly born human babies are unable to recognise partly occluded objects. By the age of four months they can tell that a partly hidden object is, in fact, a whole, single object as long as there is similar movement of both of the visible parts (e.g. a dog behind a tree trunk that is shaking both its head and tail), but at this age they cannot recognise a partly hidden stationary object. Only later does this ability develop in humans. One might ask why a young chick can recognise partly hidden stationary objects, whereas young humans cannot. A likely explanation is that chicks are precocial animals that arc already quite well developed by the time that they hatch. By contrast, the human is far less developed at birth.

Mental representations may also occur for sounds. As discussed in chapter 2, vervet monkeys use different calls to indicate the approach of different predators such as eagles, snakes or leopards, and other monkeys in their group respond in the appropriate manner to each of the calls. It would seem that those hearing the call have a representation, or image, of the predator in their 'minds'. Hearing the call allows them, as it were, to conjure up the image of the predator to which the particular call refers without seeing the actual predator themselves. There is no evidence that this is, in fact, the case because the monkeys may be responding to a specific and complex set of visual and auditory stimuli, although I suspect that this is not so.

Tool using

Much importance has been attached to tool using in humans and, until quite recently, tool using was considered to be a characteristic exclusive to humans and a hallmark of our superiority over other species. Indeed, the earliest evidence of stone-tool using in our ancestors was 2 million years ago. Numerous examples of tool using by animals have been reported now. The strict definition of tool using requires use of a separate object, not part of the user's body (i.e. not a beak or a claw) to make an alteration in another object. Using a hammer to crack open a nut qualifies as this sort of tool use, the hammer being the first object, the tool, and the nut being the second object, the one that is changed. As Christophe Boesch of the University of Basel, Switzerland, has observed, wild chimpanzees use rocks to crack open nuts, which they place on another stone that acts as an anvil. It seems that the chimpanzees understand the function of the hammer and anvil because they place the nut on the hardest part of the anvil before striking it with the hammer. They also vary the manner of hammering according to the quality of the nuts. The chimpanzees take a rather long time to learn to crack open nuts and, as discussed in chapter 2, mother chimpanzees have been observed teaching their offspring to do so. Learning to crack open nuts also occurs by observation of others performing the behaviour and by facilitation, because the right kinds of stones for hammering and for use as anvils are left together in the place for cracking open nuts.

Chimpanzees also use tools to 'fish' termites from their nest. In fact, they even fashion the tool that they use. They break off stalks of grass or twigs to an appropriate length and then insert them into the holes in a termites' nest. The termites grab hold of the stalk with their pincers and the chimpanzees pull out the stalk covered in termites, which are then eaten. This form of tool using occurs in several different groups of wild chimpanzees in different regions, but there are regional variations in tool use. Termite fishing is carried out by chimpanzees in some localities but not others and the same is true of nut cracking. Each form of tool using is passed on as a culture in each of these areas. There has even been a report of a chimpanzee using a stick to 'fish' a squirrel out of its hole and then eating it.

Several other forms of tool using have been seen in wild chimpanzees. These include using sponges to obtain water for drinking from inaccessible crannies and even using a toolkit to get access to honey. Brewer and McGrew reported the case of a chimpanzee that, firstly, took a large sharp-ended branch and used it to chisel a hole in the wax coating of a beehive. Next it used a smaller and thinner stick for more accurate work on the hole, and then it fashioned a green branch to about 30 cm in length and used it to puncture the seal over the honey. Finally, it extracted the honey by dipping a green vine into the hole.

So far, there have been fewer reports of wild orang-utans using tools, compared with chimpanzees, although tool use is very common in captive orang-utans. I suspect that this is because there has been far less observation of wild orang-utans than there has been of chimpanzees. However, wild orang-utans in a part of the Sumatran rainforest in Indonesia have been observed to fashion a tool to probe into holes in trees, presumably to extract insects or sap. The orang-utans selected a stick from which they stripped the leaves, then chewed it at one end and split it at the other end to form a spatula shape. The spatulate end was held in the mouth and the chewed end was hammered into the hole. Next the tool was withdrawn from the hole and the chewed end was inserted in the mouth.

Gisela Kaplan and I observed a new form of tool using in rehabilitated orang-utans in Sabah, East Malaysia. These orang-utans are fed bananas and other fruit on platforms located in the jungle. On more than one occasion we noticed an orang-utan spitting a mouthful of chewed banana flesh onto a 'plate' that it has fashioned from a number of leaves, spread like a fan. The orang-utan used the plate at a distance high up from the table, after carrying the banana in its mouth to this more secluded spot where it proceeded to eat slowly without competition from others.

In captivity or other forms of contact with humans, orang-utans imitate the way in which humans in their vicinity use tools, as discussed in chapter 2. They also use leaves to sponge up water, as do chimpanzees, and clean their teeth and ears with sticks. Gisela Kaplan and I have observed all of these forms of tool using in rehabilitated orang-utans in East Malaysia. Wild orang-utans probably do the same things but have not been observed to do so yet. There are other forms of tool using that do not fit the strict definition of tool using that I gave at the beginning of this section, but many would consider them to be tool using nevertheless. These include breaking off and throwing sticks at intruders, performed by both chimpanzees and orang-utans, as well as using leafy branches to fan away insects. Wild orang-utans have been sighted using leaves to wipe faeces from their infants' hair.

Apes in captivity have shown themselves to be capable of the kind of tool use that has been associated with early hominids (ancestors of modern humans). A captive chimpanzee was given a problem of getting food from a box tied up with string. The chimpanzee fashioned a cutting tool by striking a hammer stone against a cobblestone, thereby making sharp flakes. One of the flakes was then used to cut the string around the box. This is clearly sophisticated tool manufacture and use. The same has been observed in a captive orang-utan and in South American capuchin monkeys. Capuchins in captivity produced stone flakes by striking rock cores against hard surfaces and then used the flakes to take the flesh off bones and to cut through barriers. Captive capuchins also manufactured tools from bamboo when they were given pieces of bamboo and containers of sweet syrup that could be reached either by probing a tool into the container or by cutting it. The capuchins manufactured both probing and cutting tools from the bamboo and thus managed to eat the syrup. As Charles Westergaard and Stephen Suomi, the researchers who conducted these experiments, pointed out, the tool-making techniques of the capuchins are analogous to those that have been hypothesised for prehistoric hominids.

Of course, the cognitive steps that are involved in tool using must be considered. There might be planned or purposeful use of a tool, or a tool may come to be used purely by chance as all strategies are brought to bear on a problem. Elisabetta Visalberghi of the Instituto di

Psicologia in Rome, Italy, has studied tool-using behaviour in capuchins and claims that, unlike chimpanzees, capuchins do not use mental abilities to solve the tasks in which they use tools. Rather, she claims, they make persistent trial-and-error (unplanned) attempts using a variety of objects, one of which chances to be a tool that is used to solve the task. Thus, she concludes that, in contrast to chimpanzees and humans, capuchins never develop an understanding of the requirements of the tool tasks. However, this conclusion would not explain the examples of tool manufacture by capuchins mentioned above. The importance of Visalberghi's conclusion lies in its separation of the tool-using behaviour of humans and their closest relatives, the chimpanzees, from all other species, capuchins being New World monkeys that branched off early from the line of evolution that led to humans. Thus, tool using, once thought to be the hallmark of 'humanness', is redefined and can be extended to chimpanzees—other apes notwithstanding—but not beyond them. Tool using by monkeys, according to this position, is not the same thing as the planned and considered tool use of chimpanzees and humans.

Apes and monkeys are, however, not the only animals that use tools. A sea otter holds a rock to its chest as it floats on its back and uses this as an anvil against which to crack open shellfish. Chevalier-Skolnikoff and Liska have found that elephants in a zoo perform more than twenty different kinds of tool use, and nine types have been observed in wild elephants.

Tool using, and even tool manufacture, also occur in birds. Some species of finches on the Galápagos Islands use cactus spines to probe into crevices in order to impale insects. George Millikan and Robert Bowman of San Francisco State College, USA, have conducted a series of experiments in which they gave captive woodpecker finches from the Galápagos Islands various tools (short and long sticks, bent and straight ones) and different manipulative tasks (Fig. 3.4). They found that hungry birds used more tools to probe into crevices to obtain meal worms than ones that were not hungry. The bird would first try to get the worm with its beak, and, if it failed to reach the worm, it would take up a tool to probe for it. This suggests that taking up the tool is a deliberate act with a plan in mind but, again, a simpler stimulus–reward explanation could also be found to explain the behaviour. The woodpecker finches in these experiments were also clever enough to pull up a string hanging from a perch to obtain a meal worm tied at the end of it. They did so by taking the string in the beak and standing on each loop of the string after it had been pulled up in the beak. In fact, there are a number of species of birds that can carry out this manipulative feat, including North American crows.

Very recently, Gavin Hunt of Massey University, New Zealand, reported both manufacture and use of tools by crows to probe for insects. Hunt studied wild crows in New Caledonia and found that they manufacture two different kinds of hooked tools to help them capture prey. One kind of tool is made by choosing a twig with a hooked end, working with the bill on the hook end and then stripping the twig of its leaves and bark. The other kind is cut from pandanas leaves. The birds even stored the tools for using again and they appeared to choose the appropriate tool for a particular requirement. These two behaviours would require some forward planning, which is considered to be an aspect of consciousness, although there will need to be some well-designed experiments carried out with the crows to prove that this is really the case. Crows are particularly prone to using tools: the North American crow will even learn to use a stick to probe into a hole to push a key to get a food reward.

Tool making has also been observed in northern blue jays by Thony Jones and Alan Kamil of the University of Massachusetts, USA. The blue jays were seen to tear pieces from the pages of newspapers to use them as tools to rake food pellets that were out of direct reach of the beak in through the wire of their cages so that they could eat them. There are other examples of tool making and use in birds,

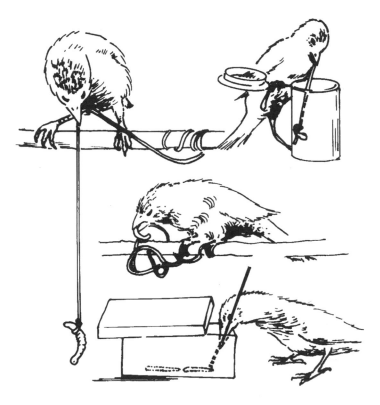

Fig. 3.4 Woodpecker finches, *Cactospiza pallida*, from the Galápagos Islands use sticks as tools to probe for meal worms. They also pull up a hanging string with a meal worm tied at the end
Source: Millikan and Bowman, 1967.

but these should serve to establish that the tool-using behaviour of birds is, as far as one can see, as sophisticated as that of primates and, indeed, early hominids.

Birds and primates in the wild have, so far, been observed to manufacture their tools only from perishable materials (the termite-fishing sticks of chimpanzees, the probing tools of orang-utans and the probing and cutting tools of crows), but stones are used as tools by wild chimpanzees to crack nuts, by otters to crack shells and also by birds to crack eggs. For example, the Egyptian vulture throws stones at ostrich eggs in order to break them and the black-breasted buzzard of Australia flies up and drops stones onto emu eggs to break them. I am not aware that anyone has studied how these egg-breaking behaviours are acquired but there is every possibility that the process is similar to nut cracking in chimpanzees. It would appear to be just as skilled.

In the case of the primates, some researchers have argued for the existence of parallel evolution in the South American capuchins and the apes. That is, tool using is thought to have arisen separately in both of these lines of evolution. The existence of tool-using behaviour in birds might be taken to suggest a third line of parallel evolution (i.e. yet another independent evolution of tool using), or it may suggest that tool-using behaviour was shared by a common ancestor of birds and all of the primates. The common ancestor idea would mean that tool using appeared very early in evolution. The parallel lines of evolution would suggest that tool using is not an unusual acquisition. Either way, the evidence goes firmly against the position that tool using is a special characteristic of humans.

What can we conclude?

In this chapter we have seen that animals are capable of doing all sorts of complex and clever things, but perhaps, as the psychologist Nicholas Humphrey said, they have clever brains but blank minds. Similarly, Nicholas Mackintosh of Cambridge University, UK, claims that we are far too inclined to attribute to animals more complex mental states than their behaviour actually warrants. He acknowledges how clever the behaviour of animals can be but prefers not to attribute to animals anything like human intelligence, or presumably consciousness.

It is true that, in certain states of mind, even humans may perform complex behaviour without being aware of what they are doing. For example, sleep walkers can negotiate stairs and even climb on roofs without falling but they are not aware that they are doing it, nor can they remember it after they wake up. Others speak whole sentences in their sleep but do not know that they are doing so. 'Blind sight' is another case of behaving without awareness. After extensive injury to the cortex of the brain, some people think that they are blind, but if they are asked to guess where an object is or what it looks like they can answer correctly. They are able to process the visual information and answer correctly without being aware that they have seen anything. Is this what the animals that I have mentioned in this chapter are doing? I think not, but many people do think so. Being intelligent is clearly a basis for consciousness but it does not prove that consciousness is present.

CHAPTER FOUR

EVOLVING A BRAIN FOR CONSCIOUSNESS

The brain is made up of nerve cells (called neurons), which conduct electrical signals and are connected with each other to form neural circuits. There are many different kinds of neurons as well as other cells, known as glial cells. Glial cells provide nutrition and structural support for the neurons and serve a number of other different functions in the brain. This is the material of the brain, out of which the mind must emerge somehow and somewhere. Can we find some aspect of brain structure or electrical activity of the neurons and their circuits that might be the material basis of consciousness? Some neuroscientists believe that this will be possible, whereas others (e.g. the late Roger Sperry of the California Institute of Technology, USA, writing in the 1980s) have argued that scientists will have to look beyond the material aspects of the brain in order to understand consciousness.

Even if Sperry is correct, it remains important for us to see whether we can explain consciousness and intelligence based on brain structure or some other measurable aspects of the cells in the brain. Perhaps conscious thinking occurs in a particular part of the brain where neurons are arranged in special ways. Perhaps we can measure some aspect of the electrical and molecular function of a neuron or of neural circuits, essential for consciousness. This would have to be a property of the neurons that is present only in the conscious state and not when the animal is sleeping.

There has been a recent renewal of interest in searching for neural mechanisms of this kind. At the 1996 Congress of Psychology held in Montreal, Canada, there was a symposium devoted to the neurophysiology of consciousness. Not surprisingly, the neurophysiologists have turned to animals in their experimental search for these mechanisms. For example, Dr R. Llinaus, of the New York University Medical Center, USA, presented a paper about consciousness and the physiological properties of neurons and their circuitry, and illustrated his points by electrophysiological recordings from animals. After his talk, a member of the audience asked whether he considered that animals have consciousness. His answer was a direct affirmative. Was this a pragmatic belief to underscore his experimental requirements to work on animals or one based on assessment of the evidence? He did not elaborate.

For those more traditional thinkers who have reserved consciousness for the human mind, the approach has been to find the explanation for consciousness in brain structure. Three main aspects of the structure of the brain have been implicated. The presence of consciousness in humans has been attributed to our larger brain size compared to all other species, to the presence of a well-developed neocortex and to the lateralisation of the brain. I will discuss each of these in turn.

Brain size and evolution

To link overall brain size to intelligence, and ultimately to consciousness in humans is, to put it mildly, a rather sweeping approach and one for which I have little affinity. It needs to be discussed, however, because increasing brain size is frequently asserted as the explanation for the evolution of human superiority.

In a very general sense, variation in brain size between different species reflects cognitive ability or intelligence. A larger brain contains more neurons, which transmit information in the form of electrical signals. The electrical circuits so formed are used to process information, and therefore a larger brain can handle more information. Neurons also play an essential role in memory formation. A cascade of molecular changes occurs in neurons when a memory is laid down. It is possible that a brain with more neurons might form more memories or more detailed memories, although we do not know exactly how this might occur.

It is not the size of the brain alone that counts. If this were so, elephants would be much more intelligent than humans. We must not consider brain size without taking into account body size. Species with bigger bodies have proportionately larger brains because a certain amount of the brain must be given over to controlling muscular movement and maintaining physiological functioning. A bigger body has a larger mass of muscles to control and a larger surface area to monitor. Small fish have small brains and large fish have large brains, and there is a direct relationship between brain weight and body weight across all of the species of fish. If brain weight is plotted against body weight, each on a log scale, for a large number of species of teleost (bony) fish, a straight-line relationship is found (i.e. as body weight increases so does brain weight in a systematic way; see Fig. 4.1). The same relationship will emerge for other groups of animals if we plot them likewise. A straight-line relationship exists for reptiles, birds, lower mammals and primates.

For each group of animals, the slope of the line plotted is less than one, which means that, although brain weight increases with body weight, it does not quite keep up. This probably does not mean that heavier species have a lesser amount of brain capacity left over for doing things other than moving and monitoring their large bodies but, rather, that the efficiency of neural circuitry improves with increasing size. After all, we know that elephants have very complex cognitive abilities, as indicated by their learning capacity, long memories and tool use. As mentioned in chapter 3, Chevalier-Skolnikoff and Liska have reported over twenty different types of tool use in elephants.

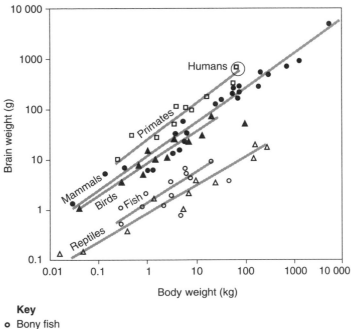

Key
- ○ Bony fish
- △ Reptiles
- ▲ Birds
- ● Non primate mammals
- ▫ Primates

Fig. 4.1 Brain weight is compared with body weight for different species of bony fish, reptiles, birds, nonprimate mammals and primates *Source:* Simplified from H.J. Jerison, 1973, *Evolution of the Brain and Intelligence.* Academic Press, New York. Also in Bonner, 1980.

There are differences between the brain-weight to body-weight ratios of animals in the different groups. Although the plotted points for fish and reptiles fall on roughly the same line, those of lower mammals and birds are on a line slightly above this, meaning that they have consistently larger brains for a given body weight. In other words, if we were to take a species of fish and a species of bird that had equivalent body weights in the adult form, the bird species would have more brain in proportion to its body than the fish species.

The line for primates is shifted yet a little further above that of birds and lower mammals. For example, a hedgehog weighing 860 grams has a brain weight of around 3.4 grams, whereas a galago, a lower primate, of the same body weight has a brain weight of around 10.3 grams. When adjusted for body size the brain weight of primates is greater than for all of the other groups, although there is still variation within the groups. We can compare the brain weight of the 860 gram galago with that of a New World primate, the squirrel monkey, weighing around only 700 grams but with a brain weight of over 20 grams. Amongst the primates, the human brain is the largest in proportion to body weight compared with all other species. Some 1.5 million years ago the human brain took an evolutionary leap forward and increased in size relative to body weight. This will be discussed in more detail later.

The order of increasing brain-weight to body-weight ratios from fish and reptiles to birds and lower mammals and then to primates and, lastly, humans reflects the order of evolution (Fig. 4.2). Amphibians evolved from fish and reptiles evolved from amphibians. Reptiles gave rise to two branches of evolution, the birds and the mammals. Primates evolved from lower mammals, and apes, which include humans, are the most recent primates to evolve. Throughout this trajectory of evolution the brain was increasing in size relative to the body. Are we at the pinnacle of this evolution? Does our large brain-weight to body-weight ratio explain our 'superior' intelligence and consciousness?

Many people think so and, in the past, some scientists have gone so far as to consider that differences in brain size between the sexes and races of humans might explain the social dominance of some groups of humans over others. One hundred years ago it was argued by researchers such as the neuroanatomist P. Broca and his colleague

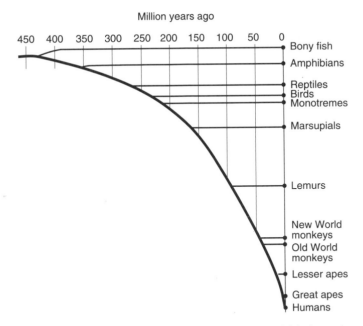

Fig. 4.2 Evolution of the vertebrates. The dates at which the various groups first appeared are based on data of DNA hybridisation (described in chapter 5). The samples for analysis were all presently living forms; hence the list of names at time zero. This evolutionary scheme is quite similar to that determined from the record of fossils.

G. Le Bon that the white male brain was larger than that of women and black people. It thus became fashionable to measure the size of the brains of eminent men after their deaths, but the weights of some were found to be so embarrassingly small that the fashion waned. Brain weight does not bear any relationship to the differences between individuals within the same group, let alone within the same species.

Brain weight is a global and gross measurement even when it is adjusted for body weight. Perhaps it might explain some of the differences in cognitive capacity between the groups of fishes and birds and so on, but within the brain there are numerous regions each specialised to perform one or some functions and not others. Only if one thinks of intelligence in a unitary way is overall brain size a consideration, and even then intelligence must depend on the neural circuitry in the many different regions of the brain and their interactions with each other. Each species tends to be uniquely adapted to survive in the environment in which it finds itself. One environment might demand certain skills for survival and another other skills. Hence, there might be many different types of intelligence. In each species, the brain regions specialised to carry out the behaviour required for survival might expand in adaptation to the particular environment. In other words, as John Krebs of Oxford University, UK, has said, cognitive capacity may occur in a number of modules or elements, each adapted for the particular environment in which the species exists. We might therefore look at the size of particular regions within the brain, rather than the whole brain itself, and see whether they correlate with specialised skills or modules of cognitive capacity.

John Krebs together with Nicola Clayton, who is now at the University of California in Davis, USA, have done just this. They have measured the size of the part of the brain involved in spatial learning in birds that store their food and in those that do not, as mentioned in chapter 3. That area of the brain is called the hippocampus, and it lies along the dorsal and midline surface of the forebrain of the bird (Fig. 4.3). They calculated the volume of the hippocampus relative to the rest of the forebrain as well as adjusting for body weight. The relative size of the hippocampus is larger in species that store and retrieve food than in species that do not do so. The demand for the storing bird to have the ability to remember where it has stored its food has been met by an enlargement of the area of the brain that processes the information used for this behaviour. In species that are required to perform other cognitive feats in order to survive, there may be an

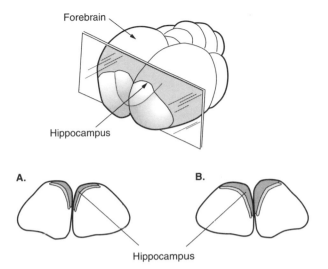

Fig. 4.3 The brain of a bird is shown with a slice through the region of the forebrain that contains the hippocampus. Slices at this angle give cross sections that reveal what is inside the forebrain. In a cross section taken from a species that does not store food (A), the hippocampus is much smaller than in one taken from a species that does store food. Based on Krebs et al. 1996.

expansion of regions of the brain other than the hippocampus.

Let me give another example of enlargement of specific regions of the brain for specialised behaviour. Only certain birds sing (pigeons, chickens and other Galliformes do not sing) and in the forebrain of birds that sing there are a number of distinct clusters of neurons, called nuclei (not to be confused with the nuclei inside cells) that control singing behaviour. In fact, there is an intricate system of interconnected nuclei that are involved in both the perception and recognition of song as well as the output of singing behaviour (Fig. 4.4). Fernando Nottebohm and

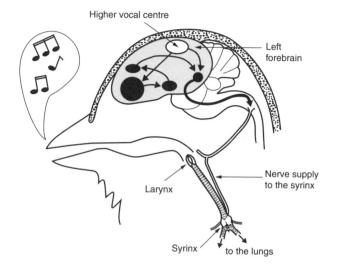

Fig. 4.4 The left hemisphere of a canary's brain showing a collection of nuclei that are involved in singing. HVC is the higher vocal centre. The song is produced by the syrinx (not the larynx, as in mammals), which is located at the place where the air passage to (and from) the lungs branches into two *Source:* Adapted from Nottebohm, 1989.

his colleagues at Rockefeller University in New York, USA, have discovered that in the spring, when song birds defend territory and sing, these nuclei increase in size by the addition of new neurons. That is, they enlarge when they are needed and shrink at other times. The ability to make new neurons like this is a rather remarkable ability of the avian brain, not present in mammalian species.

One of the nuclei involved in both perception and production of song is called the higher vocal centre. In 1993 DeVoogd, Krebs and their colleagues found that the size of this nucleus in different species of song birds correlates with the complexity of song in the various species.

Thus, the size of this nucleus appears to reflect its functional capacity.

Even within a species, there may be a relationship between the size of a particular region of the individual's brain and that individual's capacity to perform a specialised behaviour. Nottebohm has found that there is some degree of correlation between the size of the higher vocal centre, and the size of the individual songbird's repertoire. Canaries add to their song each year and individuals sing specific songs. Nottebohm analysed the canary's songs by breaking them down into phrases, syllables and elements and thus he was able to rank songs according to their complexity. This ranking had a positive relationship to the size of the higher vocal centre, although there was a reasonable amount of variation in the data. Even if the relationship between singing behaviour and nucleus size is not perfectly consistent, indicating that other factors must influence it, the results suggest that the size of a specialised region of the brain may reflect an individual's capacity to perform the behaviour associated with this brain region. For example, two birds of the same species might have the same total brain weight (appropriately adjusted for body weight) but one may have a larger relative size of the song nuclei and sing a more complex and varied song, whereas the other may have another part of its brain enlarged and perform better in the behaviour controlled by this brain region. So far, there have been no experimental studies showing this, but it is a reasonable prediction to make.

There is another important factor that we must take into account when we consider brain size. The overall size of the brain is affected by experience, and the size of regions of the brain is affected by performance of the behaviour associated with a particular region. Considering the overall brain size first, Marion Diamond at the University of California, Berkeley, USA, has demonstrated that rats raised in an enriched environment develop a larger brain than those kept in an impoverished environment. The enriched environment was one with other rats present and toys to play with, and the impoverished environment was in isolation from other rats and in a standard, boring laboratory cage. The size of the brain of the rats in the enriched condition increased by expansion of the thickness of a region of the brain called the cortex. The number of connections between the neurons increased and the sizes of the points of contact between the neurons (the synapses) increased by a remarkable 40 per cent. That is, enrichment caused an increase in the amount of connectivity between neurons in the cortex, and the cognitive capacity of the rats changed along with this. The rats from the enriched environment had superior abilities in finding their way through mazes to find food. These changes occurred after as little as thirty days in the enriched environment and in both young and old rats. Thus, cortex size is not a fixed aspect of an individual but varies with experience.

A similar dependence of size on experience has been found for the hippocampus in the food-storing birds. The opportunity to store food is essential for enlargement of the hippocampus in the food-storing birds. Krebs and Clayton prevented marshtits from being able to store food by feeding them on powdered food. Later, at various ages, they were given pieces of food which could be stored in artificial trees inside a room. Following the storing experience, and at all of the ages, the volume of the hippocampus increased. Two processes appear to have led to the increase in size. More neurons are formed and fewer are lost by natural attrition. If marshtits are completely prevented from storing food, the volume of the hippocampus decreases because the neurons in the hippocampus are not replaced as fast as they die.

These recent findings show us that the brain is in constant interaction with the environment and that use or disuse affects its size and neural circuitry. Of course, here we are talking about effects within a species. As far as we know, it is not possible to make one species equivalent to another through experience, even in the case of closely related species. Krebs and Clayton have investigated this by giving nonstoring birds the opportunity to use spatial memory to retrieve food in the laboratory. The experimenters had strategically hidden food inside small holes in artificial trees and the birds were released one at a time into the room to retrieve the food. Although nonstorers will not store they will retrieve, and they might make use of spatial abilities to remember where the food is hidden. They were compared with a storing species that, until the time of the experiment, had been deprived of the opportunity to store or retrieve. Therefore, the hippocampus in both species would have been small at the commencement of the experiment, roughly the same size relative to the rest of the forebrain in both species. After the birds had the opportunity to retrieve food, the hippocampus of the storing species increased in size relative to the rest of the forebrain but that of the nonstoring species remained small. As Krebs has said, we cannot be sure that the nonstorers did, in fact, use spatial memory in the task. They might have used other cues, such as details of the pattern or colour of the area surrounding the hole, to remember the location of the food. In fact, other experiments have shown that nonstorers do, in fact, have more tendency to rely on colour rather than spatial cues to find food. Use of a nonspatial strategy would have prejudiced the results of this experiment looking at the effects of retrieving food on hippocampal size because attention to cues other than spatial ones would have utilised other regions of the brain than the hippocampus. Nevertheless, we can conclude that the same environmental demand has not changed the hippocampus of the nonstoring species to become like that of the storing species. Thus we can consider large differences between species from an evolutionary point of view as characteristic of the species, even though experience might influence their development. This may be the case for most comparisons involving large differences in the size of various regions of the brain and species differences in the overall organisation of the brain. However, we must always keep in mind the influences of the environment on the development of the brain.

101

What might differences in the overall size of the brain mean at a functional level? A bigger brain with more neurons and more connections between neurons may function more efficiently or more 'intelligently' than one with fewer neurons but this is not necessarily so. It depends on how the neurons are connected to each other and possibly on many other factors that we do not yet know about. There are other cells in the brain, the glial cells mentioned earlier and, as quite recently discovered, they even have some part to play in the electrical activity of the brain. The number and distribution of the various glial cells may influence how a brain functions. Marion Diamond looked at a small part of Einstein's brain, preserved after his death, and found that it had relatively more glial cells as a ratio to neurons than the average human brain!

The assumption that 'bigger is better' is the basis of most theories about the evolution of the human brain made by anthropologists and many biologists. While this may have some validity when one is comparing the brains of closely related species, for example chimpanzees and humans, recent knowledge about the avian brain certainly throws the assumption that bigger is always better into doubt. As discussed in chapter 3, birds can perform problem-solving tasks and other complex cognitive tasks just as well as can primates, despite the fact that birds have very much smaller brains and, of more importance, a lower ratio of brain weight to body weight. The brains of birds are made up of neurons and glial cells the same as the mammalian brain but are organised quite differently. There is another major difference between avian and mammalian brains: new neurons can be made in the adult avian brain but not in the adult mammalian brain. The mammalian brain, the human brain being one of these, makes new neurons (and glial cells) when it is growing before birth and for a time after birth, but after this growth phase no new neurons can be formed, even to repair damage. There might be a little residual ability for the adult mammalian brain to form neurons, as Arthur Scheibel at the University of California, USA, did chance to see a neuron dividing to form a new one in a preserved specimen of a cat brain, but this ability is negligible. No dividing neurons have ever been seen in the adult brain of primates.

No one knows why adult birds retain the ability to make new neurons whereas mammals do not, but Fernando Nottebohm has made a plausible suggestion about what function this ability might serve in birds. A bird with a heavy brain relative to its body weight would have more difficulty in flying. Brain tissue is very heavy, and a heavy head, so to speak, might make a bird nose dive or force it to fly in a less aerodynamically streamlined posture. Therefore, Nottebohm suggests, the bird may vary the sizes of different parts of the brain at different times of the year as they are required. As he has shown, the sizes of the song nuclei in the forebrain increase during the breeding season when singing is required. Presumably, at the same time the sizes of other brain regions might shrink so that the increased size of the song nuclei might be accommodated within the skull. Of course, there might be other means of accommodation such as diminishing the volume of the ventricles (fluid-filled spaces) in the forebrain or decreasing the fluid-filled gaps between cells in the brain. So far no one has compared the size changes in the song nuclei with other regions of the same brain.

However, many song birds migrate, and very recently John Krebs has found some evidence that the experience of migration increases the size of the hippocampus in the European garden warbler and in a species of finch. As migration demands highly developed spatial abilities used by the bird in navigation, this result is entirely consistent with Krebs' and Clayton's earlier work on the hippocampus. For our present consideration, we may take the garden warbler's life history one step further and propose that once it has arrived, with its enlarged hippocampus, at the site where it will breed, its song nuclei will enlarge as it begins to sing to advertise its sexual attractiveness and advertise its territory. Depending on the spatial abilities that the bird must use to monitor its territory, the hippocampus may stay enlarged or regress in size. If the latter occurs, there might be time sharing of the different brain regions, and in this way the bird can keep its overall brain size smaller at any one time of the year. In other words, by juggling one area against another it might keep brain weight at an optimal low level.

The present findings point to this possibility but there is much more research needed to prove or disprove it. However, we can say definitely that it is invalid to use brain size as an index of comparative 'intelligence' between birds and mammals. I want to emphasise the special abilities of birds because they are usually left out in debates about the evolution of consciousness. There is an underlying linear concept of the evolution of consciousness along the mammalian line, reaching its highest form in humans. Having diverged earlier from the mammalian line of evolution, birds are almost always ignored. But they have developed cognitive abilities comparable to those of mammals, even primates, using different neural circuitry and special abilities to form new brain cells.

Although adult mammals cannot increase the size of regions of their brains by making new neurons, they can, as mentioned previously, increase the size and number of connections between neurons depending on experience, and this expands the size of the particular brain region. Thus, even in mammals, the size of various brain regions is not fixed and is not exclusively a result of biological predestination. Instead, it is determined by the interaction between biological events and environmental factors acting throughout the life span.

In early life the brain of mammals, as well as other species, is particularly dependent on environmental stimulation and experience. If, for example, normal visual experience does not occur, the region of the brain (the visual cortex) that normally processes visual information is taken over by auditory neurons (which respond to sound) which invade it from a nearby area of the brain. Apparently, the brain maximises the kind of processing that it has to carry out in early life. It adapts quite remarkably and this experience-dependent development affects brain function for the rest of the life span.

The study of environmental influences on the development of the brain is a major focus of the field of neurobiology, but anthropologists and psychologists have paid little attention to these new discoveries. When the abilities of different species are compared much more consideration should be given to the effect of experience on brain size and organisation. The problem-solving abilities of animals raised in impoverished conditions in animal houses or laboratories are often compared with those of humans. We do not know how much of the apparent superiority of humans over chimpanzees, for example, results from our vastly enriched experience compared with the experience of the laboratory-confined chimpanzees to which we have been compared, and how much can be attributed to the genetic endowment of our species. Yet almost always the differences found are attributed to genetic causes alone. They are seen as immutable hallmarks of the different species. If any cognitive gap exists between humans and apes, then it has surely been widened by all of the laboratory-based studies conducted so far. I suggest that we are inclined to be less critical of experimental design and the interpretation of the data when the results seem to show what we desire: human superiority.

There are even problems in comparing the cognitive abilities of one species raised in captivity with those of another species also raised in captivity, because species vary in their adaptability to captivity and to isolated or group living. Orang-utans, for instance, are less active and apparently more depressed in zoos than are chimpanzees. Presumably the same occurs in captivity in the laboratory. These differences in adaptation to captivity are, perhaps, characteristic of the two species, but measured differences in cognitive ability may be merely the outcome of the effects

105

of captivity rather than themselves being characteristic of the species.

There may also be individual differences in adaptation depending on past experience or other factors. Marion Diamond has suggested that age may be a factor in this, as we know it is in humans. She suggested that old rats may fare better in isolation, whereas younger ones do better when living in groups. Rarely, if ever, are species and individual differences such as these taken into account when species are compared in terms of cognition or other behaviours. In fact, very often, data collected from one, two or a few members of a primate species are taken as representative of the entire species. The sign-language abilities of the few chimpanzees or orang-utans so trained are interpreted as indicative of their entire species, although we would never do likewise with data collected from a few humans. We recognise that humans vary enormously but, as discussed in chapter 2, we do not attribute the same variability to individuals of other species.

To return to brain size and brain organisation, these also vary with experience. By emphasising this, I do not want to discard evolutionary theories of brain size and cognition completely; rather, I wish to raise a considered element of doubt about making definite statements linking brain size to cognitive ability, intelligence or consciousness. All too often, we see diagrams of animal brains ordered from small to large as representing intelligence or, to use a presently more acceptable term, cognitive complexity (see Eccles, 1989, diagrams 37–39A; listed in the section on further reading). The size of the whole brain and of the cortex, with increasing convolutions on its surface (called fissures), is the only criterion taken into consideration. With their small brains, which have few, if any, convolutions on the surface, birds fall close to the bottom of this hierarchy, but this ranking does not match their cognitive abilities.

The avian brain has solved its cognitive demands in a way quite different from that of the mammalian brain, and its small size indicates nothing of its cognitive complexity. It may well be that the size of a particular, specific region of the brain correlates with the complexity of its specific behavioural function, as mentioned previously, but total brain size does not indicate a great deal about overall cognitive capacity, or 'intelligence'.

Evolution of the neocortex/isocortex

Mammals evolved from reptiles over 200 million years ago and with them emerged a new layer in the cerebral hemispheres of the brain. The new layer is known as the neocortex; more recently, it has been termed the isocortex. I will keep to the older name of neocortex because it is more familiar. The neocortex became layered on top of the more primitive paleocortex, also called the allocortex (see Fig. 4.5). The exact origin of the neocortex is disputed but it appears that even the earliest mammals had six different layers of nerve cells within the neocortex. With the further evolution of mammals the neocortex expanded in size relative to the rest of the brain, and it appears to have done so many times over to give rise to different lines of mammals with different organisations of the cortex (meaning the whole cortex, paleocortex plus neocortex). In mammals with large brains the neocortex is expanded relative to the rest of the brain and the neuronal connections in the cortex are more complex, allowing more complex processing of information. The expansion of size of the neocortex was mainly along its surface rather than its thickness, and thus the surface of the cortex became more convoluted and crinkled (i.e. with more fissures or crevices; see Fig. 4.6). During the evolution of mammals, the area of the surface of the neocortex increased much more than a thousand-fold with no comparable increase in thickness. The surface area of the neocortex of a macaque monkey is one hundred times greater than that of a mouse. The relative size of the neocortex is largest in humans, a

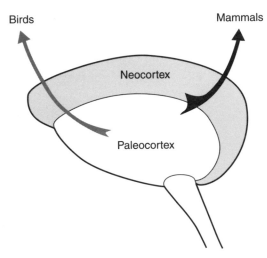

Fig. 4.5 The paleocortex means the 'old cortex'. It evolved first and, later in evolution, the neocortex was layered over it. Reptiles do not have a neocortex and nor do birds. Birds evolved more complex brains by elaborating the paleocortex: their forebrain is paleocortex. The neocortex evolved with mammals and it expanded in size as evolution proceeded.

thousand-fold greater than the surface area of the neocortex of the macaque monkey.

One can ask what factors in the environment might have influenced the evolution of a larger neocortex in some species compared with that of others. Among the non-human primates, monkeys and apes, it seems that diet and social relationships were significant factors in selecting for different sizes of the neocortex in different species. Toshiyuki Sawaguchi of the Primate Research Institute in Kyoto, Japan, divided a large number of nonhuman primates into different groups according to their diet, their habitat and their social structure, and measured the volume of the neocortex relative to the volume of the rest of the brain. By looking at the relative size of the neocortex, it

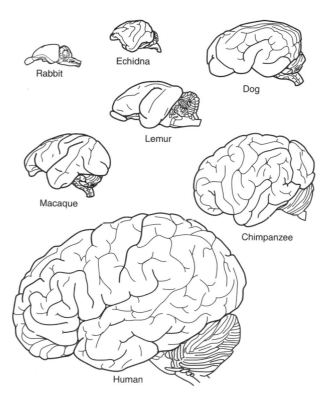

Fig. 4.6 Brains of various mammalian species showing increasing amounts of convolution of the neocortex as it increases in size.

was possible to control for variations in overall brain size that would vary with body size, itself related to diet and other factors. Thus, Sawaguchi was not interested in total brain size, adjusted for body weight, but in the way that the brain might have become organised, the expansion of one region relative to the others. The findings were very interesting. Those primates that feed primarily on fruit, although they might take some insects and leaves, were found to have higher relative volumes of the neocortex than primates that feed predominantly on leaves. This might, perhaps, be explained by the fact that fruit-eaters have to search for their food, which is usually rather sparsely distributed, a fruiting tree occurring here and there or in small clumps, whereas leaf-eaters find their food more evenly distributed. Also, fruit ripens at only certain times of the year and fruit-eaters must remember when that is. For example, orang-utans are known to visit their favourite fruiting trees only when the fruit is ripening: they remember when that is and do not need to keep returning to see if the fruit is ripe. Fruit-eaters, therefore, rely on well-developed abilities to form and remember spatial and temporal maps of their environment. These abilities might be achieved by having a large neocortex relative to the rest of the brain. However, there is a problem here because mammals process spatial information in the hippocampus (as do birds) and the hippocampus is not in the neocortex. Also, as mentioned previously for grazing animals, leaf-eaters can be quite selective in their diets.

Social structure also influenced the relative size of the neocortex. The polygamous species (ones in which males had many female partners) had significantly larger relative neocortex volumes than monogynous species (ones that formed single male–female partnerships). It is not at all clear how having more partners might influence the size of the neocortex, but the latter was also influenced by the size of the primate's social group (i.e. troop size). The larger the troop size, the larger was the relative size of the neocortex. Sawaguchi suggested that this relationship might be explained by individuals in larger troops having to remember more faces, vocalisations and behavioural characteristics of their troop members. Of course, all of these relationships do not tell us directly what the causal factors are. We can only speculate and should remember that the influences could be indirect, caused by some other factor that goes along with eating fruit or being in a large troop, such as encountering different kinds of predators depending

110

on where food is found and on more or less protection depending on troop size. Even though we cannot say conclusively what was the exact factor that led to the expansion of the neocortex, these calculations show that some aspects of the environment lead to the selection of species with different relative sizes of the neocortex.

Along with the neocortex, an entirely new structure evolved in the cerebral hemispheres of mammals. That structure is a large tract of nerves connecting the two cerebral hemispheres to each other and it is called the corpus callosum. This tract is not present in reptilian or avian brains, which have a number of much smaller tracts connecting each side of the brain. The size of the corpus callosum, relative to the rest of the brain, is largest in humans. Thus, humans have more neocortex and more connections between the separate neocortical regions of the left and right hemispheres. The corpus callosum appears to have an important role in preventing the left and right hemispheres from both carrying out the same function, that is, from duplicating functions. This appears to be possible because the corpus callosum links areas in one hemisphere to their equivalent areas in the other hemisphere, thereby allowing inhibition by an area in one hemisphere of its equivalent in the other hemisphere. This would generate lateralisation of the hemispheres (each hemisphere carrying out a different set of functions), which is discussed in the next section.

It is widely assumed that the evolution of the neocortex was associated with the evolution of intelligence and, ultimately, consciousness. In their book entitled *Neocortical Development*, written in 1991, Bayer and Altman state: 'It is widely assumed that the evolutionary growth of mental life that reaches its zenith in humans is attributable to the progressive expansion and elaboration of the neocortex' (Bayer and Altman, 1991).

The following quote is in a similar vein: ' . . . comparative neurobiology is an integral part of attempts to understand the functional organization of the neocortex and, ultimately, the evolution of more complex functions that are generated by the neocortex, such as perception, cognition and consciousness' (Krubitzer, 1995).

These are but two quotes on a common theme amongst neurobiologists, who specialise in the study of neurons, other cells in the brain and brain structure. The pitfall for these scientists is that, unfortunately, they have little familiarity with the study of animal behaviour or comparative psychology. Their knowledge of the brain itself is not matched by knowledge of the behaviour of the animals in question. Before sensible relationships can be established between brain organisation on the one hand and behaviour on the other hand, scientists need to be well versed in both fields. I stress this because perception, cognition and consciousness can be measured only in terms of behaviour and we want to be able to discover whether animals have consciousness. The statement by Krubitzer is based on the assumption that consciousness evolved only in the mammalian line.

Although the neocortex might have provided mammals with the neural substrate (i.e. neural circuits and structures) required for intelligence and consciousness, *without* a neocortex birds have complex cognitive abilities that rival those of species *with* the neocortex. The hypothesised association of the neocortex and consciousness is generated from a human-centred position. Only in humans, it is assumed, has the neocortex become elaborate enough to give rise to consciousness.

Evolving from a reptilian ancestor along a branch of evolution separate from that of mammals, birds have acquired cognitive abilities using different regions of the brain and different neural circuits. As said before, the structure of the avian brain is quite different from that of the mammalian brain. Recognition of this should tell us that the neocortex might not be essential for intelligence and cognition but, as we have seen, birds have usually been ignored or underestimated by the scientists who have written about the evolution of intelligence and consciousness.

Sir John Eccles, formerly of the Australian National University and winner of the Nobel Prize for his discoveries about the electrophysiology of neurons, has developed a hypothesis about the evolution of consciousness based on the presence of certain cells and circuits in the neocortex. In the neocortex there are neurons of a particular shape known as pyramidal cells and these are clustered into bundles called dendrons (Fig. 4.7). As far as can be deduced from modern brains in various species, the dendrons first appeared in the brain 200 million years ago in the first, primitive mammalian neocortex. There are about 40 million of these dendrons in the human neocortex. Eccles has hypothesised that the dendrons are essential for consciousness. He speculates that electrical activity in the dendrons interacts with the 'world of the mind' to produce what he calls units of consciousness, or psychons. Thus, he ties consciousness to a particular cell type, on the assumption that only in mammals did consciousness evolve. As the pyramidal cells of the neocortex are structurally very complex and have a great many connections, they are a good starting point for the structural correlate of consciousness, but they probably do not play an exclusive role in the mechanisms underlying conscious thought. Besides, it would be difficult, if not impossible, to test the hypothesis that the dendrons might be the location of conscious mental processes. Eccles mentions that birds show insightful behaviour and calls for further examination of a part of the avian brain, the Wulst, to see if the neurons there might have something in common with those of the neocortex of mammals. Birds would provide useful comparison to test Eccles' mammalian-centred hypothesis but, even if they do have neuronal circuitry similar to that of mammals, it would not prove that those particular circuits generate consciousness.

More recently a subregion of the neocortex, the prefrontal cortex, located in the frontal lobes, has been designated a special role in human consciousness. The prefrontal cortex occupies about one quarter of the human

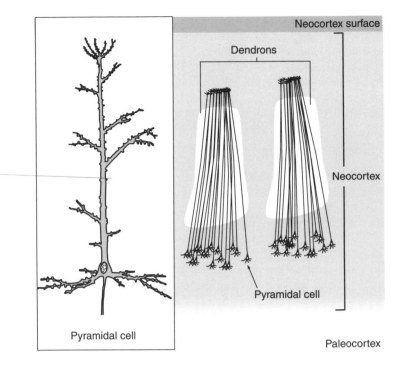

Fig. 4.7 Pyramidal nerve cells in the mammalian neocortex. A single pyramidal cell is shown on the left. These cells are interconnected in groups, called dendrons, as shown on the right *Source:* After Eccles, 1992.

neocortex, an apparent advance on the great apes which have a prefrontal cortex occupying only about 14 per cent of their neocortex. Some neuroscientists call the prefrontal cortex the 'command headquarters', and recently they have found that it has a particular form of synchronous electrical activity, known as theta rhythm, when a person is in deep thought. Insightful and self-reflective thinking in humans has been attributed to the prefrontal cortex. By implication, animals, including the great apes, may be said to lack insight and reflection, or to manifest it in a less-developed form. As a flow on from this hypothesis there has been a suggestion that autistic individuals, who are said to show little understanding of their own mental states or those of other individuals, may have impaired functioning of the prefrontal cortex, although there is insufficient evidence to substantiate this claim. Moreover, the association of consciousness with an area of the brain that happens to be larger in humans echoes the earlier arguments about bigger being better. I have already discussed the contrary evidence relating larger brain size to higher intelligence and consciousness and the same criticism could be applied to the hypothesised link between the size of the prefrontal cortex and consciousness.

A leap in neocortical size with humans?

In 1995, Barbara Finlay and Richard Darlington, writing in the journal *Science*, proposed a model that might explain the accelerated expansion of the neocortex in the evolution of mammalian brains, the human brain being at the top of an exponential increase in the size of this important region of the brain. Recognising that each species is subjected to the forces of natural selection that lead to it optimising its behaviour in a particular environment, they asked what changes might take place in the brain to allow a species to develop a particular specialised behaviour, controlled by a particular localised region of the brain. For example, as discussed earlier in this chapter, birds that use spatial information to store their food and find it again have an enlarged hippocampus. This is a special adaptation to their particular environment, the birds storing food when it is abundant and retrieving it when it is scarce. Mammals that store food (e.g. squirrels) likewise have an enlarged hippocampal region of the brain. Animals that have hands and can use them to catch prey or to manipulate objects have enlargement of the part of the neocortex that deals with the sense of touch from the hands. This region of the

115

cortex is known as the primary somatosensory area. The American raccoon, for example, has hands that it uses in catching its prey and, compared with its nearest relatives, it has a much enlarged primary somatosensory area, and a larger portion of this area is devoted to processing information received from the hands than it is in other species. In fact, the information about touch is sent from the hand in a consistent arrangement so that there is a map of the individual digits of the hand on this area of the brain. How is the enlargement of an area specialised to perform a particular adaptive function achieved, and what happens to the other brain structures when this one region expands to make an adaptation to a particular environment?

Finlay and Darlington wondered whether adaptation of a species to perform a special behaviour in a particular environment might have led to the expansion of only the region (or regions) of the brain needed for that particular behaviour or whether other areas increased along with it. They reasoned that, amongst the mammals at least, the latter may be true because, when a brain is developing, it makes new neurons in a particular order, and this order is almost identical in all mammalian species. To make one area of the brain larger and keep the same order for making new neurons, all regions that develop at the same time and after the required region would have to enlarge along with it. If this is what happens, the selection of one specialised ability (e.g. hands to catch prey) would lead to an expanded capacity to perform other specialised functions. There are some examples that seem to support this proposition. The Australian striped possum (*Dactylopsila trivirgata*) has a special adaptation for its mode of feeding in the canopy of the rainforest: it has one digit longer than the others and it can use this digit to get insects out of holes in trees. It also has the largest brain, corrected for its body weight, of all marsupials. Thus, its adaptation of a special digit may have led to an overall increase in brain size, not just an increase in the size of the brain region controlling the digit itself.

Put in other words, acquiring one special ability may enhance the brain's capacity for performing many other special functions. Thus, according to this proposal, when the hooves of ungulates (horses, donkeys, etc.) evolved first to paws (of rats, cats, etc.) and then to the hands of primates, not only did the region of the neocortex used to control the forelimbs expand in size but so did the entire neocortex. The adaptation made might have been specifically to evolve hands and the ability to use them to manipulate objects, but many other abilities went along with this acquisition of the new behaviour.

Finlay and Darlington measured the sizes of different regions of the brain of a large number of mammalian species living in different environments and plotted the size of each region against the total brain size (see Fig. 4.8). They found that, as total brain size increases, the size of the neocortex, in its entirety, expands relative to all of the other regions of the brain. The size of the neocortex increases at a faster rate than the size of the other brain regions (e.g. the cerebellum, diencephalon and paleocortex). In fact, the neocortex expands exponentially compared with the other regions of the brain. To cite an example given by Finlay and Darlington, the brain of the smallest shrew is some 20 000 times smaller than the human brain, whereas its neocortex is more than 100 000 times smaller. Of course, body weight has to be taken into account, but let us compare two species of comparable body weight: the insect-eating tenrec (a hedgehog-like mammal of Madagascar) has a brain that is ten times smaller than that of a squirrel monkey, but a neocortex that is sixty times smaller than that of the squirrel monkey.

The human neocortex is at the top of the exponential curve. Perhaps the human acquisition of some specific behaviours such as walking in a more upright posture and use of the hands in making tools led to an exponential increase in the size of the neocortex. Thus, along with these adaptations, we might have acquired the increased brain capacity for thinking, for consciousness.

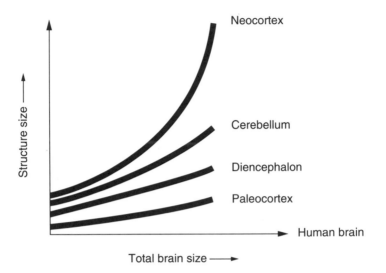

Fig. 4.8 The sizes of four different structures in the brain are compared with the total size of the brain. The data have been collected from various different species and lines have been drawn to connect them all, instead of plotting single dots for each sample measured. Note that as brain size increases, the size of the neocortex increases in an accelerating fashion (i.e. exponentially), whereas the size of the other structures increases in a more linear fashion. Therefore, with increasing brain size, the neocortex makes an increasing contribution to the total volume of the brain *Source:* Adapted from Finlay and Darlington, 1995.

This hypothesis could explain the evolutionary leap forward in the human brain, occurring 1.5 to 2 million years ago (to be discussed further in chapter 5). It should, however, apply to other branches of evolution and it could be tested for avian species. Do the food-storing species have other special abilities that evolved along with their ability to store food using their larger hippocampus? Do owls, which are perfectly adapted to searching for their food in the dark of night using specialised abilities for locating the source of sounds made by their prey, have other special abilities that they acquired with this specialisation? Indeed, are they more intelligent because of this and, to go further, might they have acquired consciousness as a consequence? These are extremely interesting questions, but at this time they cannot be answered.

Lateralisation of the brain

Lateralisation of the brain refers to specialisation of the two hemispheres of the brain to carry out different functions, to process different sorts of information and to control different behaviours. For example, in the majority of humans speech is controlled by the left hemisphere and the perceptual processes that allow us to understand language are also located in that hemisphere. The left hemisphere controls the right hand and, in most people, the left hemisphere is used to control writing and many other acts that are performed by the right hand. The right hemisphere in humans is involved with emotional behaviour, particularly negative emotions such as fear and discontent. For this reason, the facial expressions that signal these emotions are expressed more strongly on the left side of the face. On the other hand, when most people speak, the right side of the mouth opens wider and sooner than the left side. The right hemisphere is also involved in determining spatial locations of objects and thus controls our ability to find our way using maps.

These functional lateralisations of the hemispheres in humans are matched by structural asymmetries in the brain. The Sylvian fissure that runs between the two major language and speech areas in the left hemisphere is longer than its equivalent in the right hemisphere (Fig. 4.9A to 4.9C). The back part of the left hemisphere, the occipital cortex which is used for vision, is larger than the same region of the right hemisphere and the reverse is the case

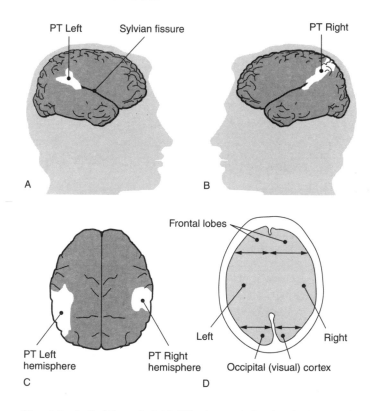

Fig. 4.9 Left (A) and right (B) views of the hemispheres of the human brain showing asymmetry in the regions used for speech: PT, planum temporale. Two views looking down from the top of the brain are also shown. C is a view of the surface showing the asymmetry in the PT regions of the left and right hemispheres and D is a section through the brain showing that the left occipital lobe is larger than the right and the right frontal lobe is larger than the left *Source:* Adapted from H. Steinmetz, 1996, *Neuroscience and Biobehavioral Reviews*, 20, 587–591.

for the front parts of the hemispheres, known as the frontal lobes (Fig. 4.9D).

Lateralisation of the human brain was discovered more than one hundred years ago when it was noticed that people who had suffered a stroke leaving them paralysed on the right side of the body suffered from aphasia, loss of the ability to speak, whereas those that had paralysis on the left side of the body had no detectable deficits in their speech. The aphasia followed from damage to the left hemisphere, which controls the right side of the body. The specific region of the damage affecting speech surrounded the Sylvian fissure of the left hemisphere.

Knowledge of lateralisation in the human brain advanced considerably with the research of Roger Sperry, who studied 'split-brain' patients, ones who had had the corpus callosum connecting the hemispheres sectioned because they suffered from severe epilepsy. This operation prevents information from being transferred directly from one hemisphere to the other. When such a subject looks straight ahead at a point on a screen and then a picture is flashed, say, in the left extreme of the subject's visual field, the visual information is sent to the right hemisphere only and processed there. If the picture is flashed in the extreme right visual field, the information is sent to and processed by the left hemisphere and this means that language and speech centres are accessed. Thus, the subjects can say the names of pictures of objects flashed in the right visual field and they can also read words flashed there but, when the same images are flashed in their left visual field, they cannot do so. For example, if an image of an apple is presented in the right field, the subject can say 'apple', but that is not possible when the image of the apple is presented in the left field because the language centre in the left hemisphere cannot be accessed. In the latter case, however, the subject is able to choose an apple from a bowl of fruit to indicate what the right hemisphere has seen. Using this technique, Sperry was able to show that the left hemisphere is specialised for forms of analytical thought, including mathematical calculation, as well as for language and speech production, whereas the right hemisphere is specialised for music appreciation, spatial abilities, expression of emotions and nonverbal processing of images. These results have been confirmed by modern techniques of 'imaging' neural activity in a living, intact brain while the subject performs a certain task (see Fig. 4.10). We know now that there are some aspects of language that are processed by the right hemisphere and that the left hemisphere is used by trained musicians to analyse music. Presumably trained musicians have learnt to use different neural circuits to analyse music. Exactly which hemisphere is used by a given individual to carry out a particular task appears to depend on past experience as well as the type of information processing used.

The existence of 'split-brain' patients, who had had the corpus callosum sectioned, provided fuel for the debate on consciousness. In the 1970s two scientists, Popper and Eccles, published a dialogue about the potential paradox of these patients having two minds in one person. If conscious thought emerges from the neocortex, there is a possibility that these 'split-brain' subjects have two separate minds because each side of the brain has a neocortex. Alternatively, since the left hemisphere is by far the one most commonly used for language, perhaps consciousness resides in the left hemisphere only. This would mean that the right hemisphere is unconscious. As mentioned in chapter 1, the issue of an obligatory association between consciousness and language underscores this debate. The right hemisphere is capable of highly complex mental processes even though it cannot express them verbally. Language is, indeed, a convenient vehicle by which we can assess consciousness, but that does not mean that a nonverbal hemisphere, or for that matter a nonverbal person, necessarily lacks consciousness. To access consciousness of the right hemisphere we would be confronted with the same difficulties as in assessing whether animals have consciousness. Not surprisingly, therefore, the question of two minds in one person remains unanswered. It has, however, been observed that some 'split-brain' people experience conflicting emotions or

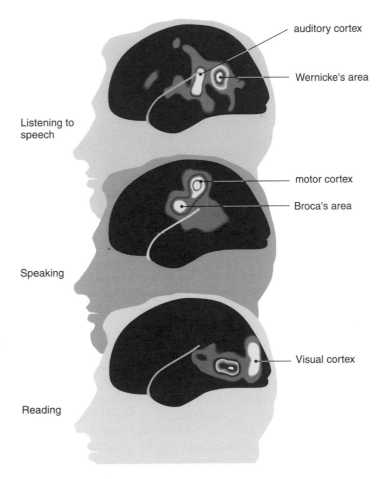

Fig. 4.10 Imaging techniques show active regions of the brain (here PET scans, meaning Positron Emission tomography). The human subject is performing different tasks and different regions of the brain are active. See Fig. 5.4 as a reference point for the regions that are active *Source:* Adapted from G.D. Fischbach, 1992, *Scientific American*, Sept., 30–31.

perform contrary acts. One subject reports opening a draw with one hand while shutting it with the other. Another reports putting one arm around her husband to greet him while pushing him away with the other hand. Are there two minds, with two moralities, in the same person? I think that these observations might suggest so. They certainly bridge the issues of consciousness in humans and animals and highlight the need to develop methods of assessing consciousness without language.

For a long time it was believed that lateralisation of the brain was a unique attribute of humans, associated with our abilities for tool use, language and consciousness. The association between these three attributes will be discussed in more detail in chapter 5 when handedness is considered also. Here we are interested in the brain functions that might depend on having a lateralised brain. Lateralisation means that there are fewer functions duplicated in each hemisphere and, thus, the capacity of the forebrain may be effectively doubled. This, it has been argued, explains the superior intelligence of humans and also our exclusive abilities of language and consciousness.

Writing in 1989, Eccles adhered to this view. To Eccles, consciousness is unique to humans and is a product of our highly developed neocortex as well as of lateralisation. He believed that all monkeys and apes have symmetrical brains, whereas asymmetry (i.e. lateralisation) evolved in humans to overcome the problem of needing more neocortex. Instead of the size of the neocortex further expanding, he suggests, functions of the neocortex were no longer duplicated on both sides of the brain. It is surprising that he held this view of lateralisation as unique to humans at the time he wrote it because there was already clear evidence that many species of animals have brain lateralisation.

As early as the early 1970s, Nottebohm and his co-workers at Rockefeller University demonstrated that there is lateralisation for the control of singing in song birds, such as canaries. First, he cut either the left or right nerve that supplies the syrinx, the organ that produces the song, situated on the airway from the lungs. Cutting the nerve on the right side had no effect on singing but cutting the nerve on the left side prevented the bird from singing. The canary performed like an actor in a silent film, going through all the motions of singing but uttering only grunts and squeaks with an occasional syllable thrown in. Later Nottebohm traced this lateralisation to the song nuclei in the brain (the ones that were discussed earlier in this chapter). Destroying the higher vocal centre on the left side of the forebrain prevented the canary from singing, but doing the same on the right side had no effect on song production. Thus, lateralisation for song production was demonstrated. The role of the left hemisphere is interesting given that bird song shares some of the aspects of human language and even involves learning: some species even learn to recognise and produce local dialects. Of course, we must remember that, if song is to be lateralised, it has a 50:50 chance of being in the left hemisphere. In fact, zebra finches have control of song lateralised to the right hemisphere, although all of the other species investigated so far use the left hemisphere.

In addition, the role of the left hemisphere in species-typical communication is highlighted by discoveries of left hemisphere specialisation for processing vocalisations in a number of species of mammals. Japanese macaques process their species-typical calls in the left hemisphere. Like humans, they show a right ear advantage for recognising vocalisations, the right ear sending its information primarily to the left hemisphere. No difference is found between ears for other, nonvocalised sounds, indicating that the lateralisation involves higher neural processing and is not simply a result of one ear hearing better than the other. Rats also show a right ear advantage for processing their species-typical vocalisations and not for other sounds. The sound that has been tested is the high-pitched, ultrasonic distress call of rat pups. If the left ear of a mother rat is blocked with wax she will, as normal, run to a loud-speaker that is emitting the sound of her pups and attempt to retrieve them, while at the same time ignoring a neutral signal being emitted from another speaker. If her right ear is blocked, she will approach both speakers at random, apparently unable to discriminate the calls of the pups from the neutral sound. Another study, by Holly Fitch and others at Rutgers University, USA, has demonstrated that male rats also have a right ear advantage for processing temporal sequences of tones, as do humans. Clinical research on humans has suggested that there may be a link between temporal processing of sounds and processing of speech sounds by the left hemisphere. Thus, children who have difficulty in learning language also have great difficulty in discriminating rapidly presented tone sequences. The same is true of people who have damage to the speech centre in the left hemisphere. The inability to handle temporal information carries over to those aspects of speech and impairs processing of language. The lateralised processing in rats might, therefore, represent a very early specialisation of the left hemisphere which later in evolution became used for language.

There appears to be nothing particularly 'human' about use of the left hemisphere to process communication signals. The same is also true for use of the left hemisphere to produce vocalisations, as we have seen in birds. Even the frog *Rana pipiens* uses the left side of its brain to make alarm calls. Also, the left side of the brain is used by male gerbils to produce the vocalisations that they make when they are courting a female. The role of the left hemisphere in vocal control and perception is a very ancient one. It is not an exclusive role and is not an entirely invariant one for all species, but it is impressively common nevertheless.

Many other brain functions are lateralised in animals. In the late 1970s I discovered that young chickens learn to discriminate food grains from small pebbles using the left side of the forebrain. Soon after that Richard Andrew of Sussex University, UK, tested chicks on the same task with a patch on the left or right eye. He found that the chicks tested with the patch on their left eye could learn to discriminate the grains from pebbles, whereas those with a patch on the right eye could not do so. This result confirms the specialisation of the left side of the forebrain for performing this task because, in birds, most of the information received by the right eye is processed by the left hemisphere and vice versa.

Following these initial studies with chicks, a large number of different functions have been found to be lateralised. To give just one more example, in chapter 2 I mentioned the studies of individual recognition in chicks tested by placing the chick in an alley way with a familiar chick at one end and an unfamiliar chick at the other end. The chick can discriminate between the familiar and unfamiliar chick and chooses to approach the familiar one. Giorgio Vallortigara, at the University of Udine, Italy, and Richard Andrew, at Sussex University, UK, continued these experiments to see if the chicks could do likewise with familiar and unfamiliar objects and then tested them with a patch over the left or right eye. Each chick was kept in a cage with a red table-tennis ball suspended about five centimetres above the floor level. The ball had a small, white horizontal strip on its equator. After a few days of becoming familiar with this ball, each chick was given a choice of the familiar ball placed at one end of the alley way and a red ball with the strip oriented vertically placed at the other end. When tested binocularly the chick notices the difference between these two stimuli and, usually, chooses to be near the familiar one. It does likewise when given a choice between the familiar ball and one with the strip oriented at 45 degrees from the horizontal. Thus, with both eyes open, both large and small differences from the familiar stimulus are detected. If, on the other hand, the chick is tested using its right eye, with a patch on the left eye, it chooses only between the horizontal and vertical orientations but not between the horizontal and the 45 degree orientation. If the right eye is patched, the chick (using its left eye) chooses between the familiar stimulus and both of the unfamiliar stimuli. In other words, small differences are noticed by the left eye and right side of the brain, whereas only larger differences are noticed by the right eye and left side of the brain. The left eye and right side of the brain of the chick are also specialised to perform spatial tasks, such as those used when searching for food.

A similar battery of lateralised brain functions has been found in rats by Victor Denenberg at the University of Connecticut, USA. Rats can also be tested monocularly because, as in birds, each eye sends its information to the opposite side of the brain. In Denenberg's laboratory rats have been tested (by P.E. Cowell and N.S. Waters) in a task requiring them to swim in a tank of water and to locate a hidden platform on which they can stand before being lifted out of the 'swim maze'. Rats can learn to locate the platform using their superior spatial abilities. They orient using cues overhead in the room and on the walls of the tank. In the monocular tests, the rats were able to perform this task well if they were using the left eye but not if they were using the right eye. The demonstrated involvement of the right hemisphere in processing spatial information is the same as in chicks and humans.

From these selected examples, it will be seen that there is no doubt that animals have strongly lateralised brains and even that the form of the lateralisation is very similar to that of humans. This is true even when we compare species with very different brains, such as birds and mammals. The corpus callosum of mammals may be important in generating lateralisation of the brain, as mentioned in the previous section, by permitting inhibition of parts of one hemisphere by their equivalents in the other hemisphere, but the corpus callosum is not essential for brain lateralisation. Birds do not have a corpus callosum, but they have strongly lateralised brains.

In animals, as in humans, there are structural as well as functional asymmetries of the brain. Chimpanzees and orang-utans have asymmetry of the Sylvian fissures, as do humans. In rats, the left visual region of the cortex is larger than its equivalent on the right side, as in humans. In birds there are asymmetries in the organisation of the neurons that transmit visual information to the forebrain.

It is now clear that the hypothesised unique relationship between brain lateralisation, language and consciousness is incorrect. It may be that consciousness could not have evolved without brain lateralisation, and this might also be true for language, but there was no simultaneous evolution of all of these attributes together.

What can we conclude?

We know of no single structure in the brain that is unique to humans, despite continual claims that have been made to this effect at one time or another. We are, it seems, always seeking to find something about the brain that might make us different from, and superior to, other species. Perhaps it is just more of everything that singles us out. Humans have the largest brain weight relative to body weight, the largest neocortex size relative to the rest of the brain, the largest prefrontal cortex and the largest corpus callosum. Perhaps these represent a special confluence of brain features, out of which consciousness emerges, or perhaps it is only a matter of degree that separates us from other species. So far, however, searching for the key to 'humanness' in brain structure has served more to dash illusions about our superiority, or simply difference, than to provide confirmation of them.

CHAPTER FIVE

EVOLUTION OF THE
HUMAN BRAIN AND MIND

A number of abilities may have come together with the evolution of humans. These proposed characteristics include standing upright on the feet (i.e. adopting a bipedal posture), the ability to perform fine manipulation with the hands, right-handedness, tool use, language, group hunting, the ability to plan ahead (intentionality) and consciousness. As we will see, many of these characteristics have also been observed in nonhuman animals. Yet, a coming together of all of these abilities may explain the appearance of the first, modern humans, *Homo sapiens*, 0.1 million years ago.

The first human-like animals, the australopithecines, appeared on earth some 4 to 6 million years ago, and some say more precisely 4.4 million years ago. Although we do not know exactly in which region of the world the transition from nonhuman apes to humans took place, the discovery of fossils that are intermediate between chimpanzees and the australopithecines in Africa suggests that this is the place where it occurred. Also, analysis of our genetic material (i.e. the genes) places us closer to the chimpanzees and gorillas of Africa than to the orang-utans of Asia. The genes are inside the nucleus of every cell in the body and they are passed on from generation to generation. They are the building blocks on which all life forms develop. Influences from the environment can radically affect what genes are expressed; as the molecular biologists say, they can affect 'the read-out from the genetic code'. Evolution occurs by changes that accumulate over time in the genes, these changes being referred to as mutations. All living species share a considerable proportion of genes (i.e. they have genes that are the same, or almost the same). This is because all of these are basic genes that need to be expressed in all life forms. They are basic for survival. These basic genes encode certain proteins that are essential to the functioning of our cells and bodies as a whole. Nevertheless, each species has a collection of genes that differs from those of other species, and those species that have evolved further apart from each other share fewer of the same genes. The fewer genes shared, the further apart are the two species in evolutionary time because, beginning from the time when they separated from each other, each species slowly accumulates different mutations of its separate genetic code. We can use these accumulated mutations as a clock to date when any two species began to evolve separately. Thus, from living animals today we can obtain information that allows us to trace their evolutionary past.

Scientists can discover how much genetic material is shared between any two species by mixing their genetic material together to see how much matching occurs between their two genetic codes. The process is called gene or DNA hybridisation. The genes are strung together in sequences like the words in a sentence, although each sentence of genes is a very long one. The genetic code of one individual is made up of many such strings of genes and thus we might consider it as a collection of sentences, some of which will be read out at different times in the individual's life and in different contexts. When gene hybridisation is carried out, the sentences describing an individual of one species are compared with those describing an individual of another species. If we hybridise the genes of a chimpanzee and a human, we find a remarkable 99 per cent similarity of the genetic code. We share slightly less than this with gorillas and 98 per cent with orang-utans. By knowing the rate at which mutations accumulate, we can date the divergence

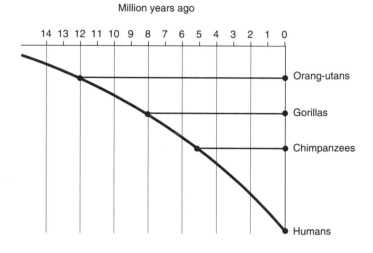

Fig. 5.1 Evolution of the hominoids, based on DNA hybridisation. This is the most accepted view, but it is not the only one. It has been suggested that humans branched off after orang-utans and that gorillas and chimpanzees evolved later on their own divergent branch. This view would explain the fact that although gorillas and chimpanzees use their knuckles when they walk, as did their ancestors, there is no anatomical evidence that the ancestors of humans were knuckle walkers. In the scheme presented in the figure, one has to assume that humans lost the knuckle-walking ability of their predecessors.

of the human line of evolution (referred to as the hominid line of evolution) from orang-utans at about 10 to 12 million years ago, from gorillas at about 8 million years ago and from chimpanzees at about 4 to 6 million years ago (Fig. 5.1). There are some inaccuracies in dating the hybridis-ation data and we should remember that the environment has a large effect in determining what genes are expressed. This may explain why orang-utans actually have more physical features in common with humans than do gorillas or chimpanzees. Yet, overall, the evidence suggests that humans are more closely related to chimpanzees. Thus, while nonhuman primates had spread out from Africa across the continents through Europe to Southeast Asia and to South America, it appears to be those that stayed in Africa that evolved into hominids.

The evolution of humans occurred at a time when Africa was cooling and becoming drier. There was a major loss of forests and an increase in grasslands, known as savannah. It has been suggested that this climatic change led to the evolution of humans that walked upright on their hind limbs (bipedally) and to a change in their feeding habits from primarily eating fruit and leaves to eating meat, for which they needed to hunt. This will be discussed in more detail later. The bipedal gait may have allowed the early humans (hominids) to move more efficiently over the grasslands in search of food or other resources.

Between the time of the first appearance of human-like animals (4.4 million years ago) and the appearance of anatomically modern humans, *Homo sapiens*, there existed a number of different species of hominids. Apparently, these species formed as a consequence of climatic changes causing fragmentation of the habitats in which they lived. This caused populations of hominids to become isolated and then to evolve along separate paths. The earliest hominid species, *Australopithecus afarensis* (which existed from 3.8 to 2.9 million years ago), is believed to have given rise to two major subdivisions of hominids: the gracile *Australopithecus africanus* (3 to 2 million years ago), which eventually led to *Homo sapiens*; and the robust *Paranthropus* or *Australopithecus robustus* and *Australopithecus boisei* (2.5 to 1 million years ago), a side branch which had become extinct by about 1 million years ago (Fig. 5.2). The gracile stock of hominids included *Homo habilis* (1.9 to 1.5 million years ago), *Homo erectus* (1.8 to 0.25 million years ago), *Homo neanderthalensis* (0.12 to 0.04 million years ago) and archaic *Homo sapiens* (0.4 to 0.09 million years ago). *Homo habilis*

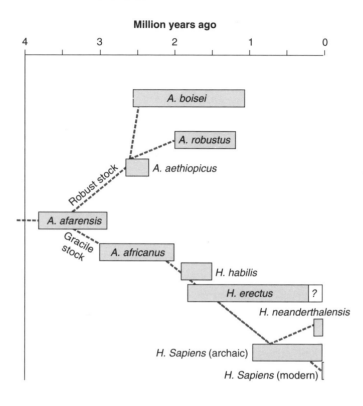

Million years ago

Fig. 5.2 The evolution of the hominids. This is a 'consensus' view and definitely not the only one that has been proposed (for alternatives see R. Gore, 1997, *National Geographic*, 191 (2), 72–97). The boxes indicate the approximate periods for which each hominid form existed. Some people prefer not to separate *A. boisei* and *A. robustus* and they assign both species to the same genus *Paranthropus*. Note the uncertainty of the end point of the period for which *Homo erectus* existed.

represented the first notable increase in brain size, relative to body size, over apes but even so its relative brain size was only half that of *Homo sapiens*.

Each of these hominid species eventually became extinct at one time or another. Some hominids survived for longer than others, probably depending on when they came into competition with later hominid forms with larger brains. There is controversy about exactly when the various hominid forms appeared and died out, and also at what time the various forms migrated from their apparent birth place in Africa to spread out over Europe, Eurasia, Australia, and so on. A recent report in the journal *Science* (written by C. Swisher of the Berkeley Geochronology Center, USA, with a number of colleagues) has made a claim, based on dating bone material, that *Homo erectus* existed in Central Java up to around 27 000 to 45 000 years ago, long after modern *Homo sapiens* had evolved. But the report needs confirmation because the age of the *Homo erectus* skulls was estimated only indirectly by measuring the ages of bovine teeth collected from the same layer of earth and from alongside the skulls. Samples of the skulls were not made available for direct dating. As critics of the report argue, the bovine teeth and the skulls may have come together by sediment drift or some more recently occurring natural phenomenon, rather than being deposited alongside each other because they lived and died at the same time.

In fact, it should be noted that all of the dates that I have cited for the various hominids are estimations only. Even when samples of skulls are available, inaccuracies result from problems involving dating the bone material, the fragmentary nature of the remains that are available and other taxonomic (classification) issues. Not only the age but also the distribution of the various hominid forms is an estimation, with similar sources of inaccuracy. However, it now appears that, although *Homo erectus* dispersed widely across the continents, it was from the *Homo erectus* population that remained in Africa that *Homo sapiens* evolved, and *Homo sapiens* then dispersed from Africa to the rest of the world. As it did so, it must have caused the extinction of the other hominids that it contacted.

None of these chronological and anatomical details are

of particular concern to us here, but they provide us with a background against which to consider brain evolution in the hominids and with a basis on which to pose the question, 'When did the human brain become the one that we know it is today?'.

Did brain capacity evolve in complete synchrony with the changes in the skeleton that are used to place the various hominids in different species, or were there steps taken by the evolving brain that occurred independently of these markers of physical evolution? Did the adoption of a bipedal posture influence the evolution of the human brain? Although we can estimate brain volume from fossilised skulls, how much does this tell us about the organisation and function of the evolving human brain? When did right-handedness and tool using emerge and were they linked to each other? When did humans begin to use language? Was it as recently as around 30 000 years ago, as William Noble and Iain Davidson have hypothesised? Can we discover anything about the consciousness of hominids from the palaeontological records? I will consider each of these proposed aspects of human evolution in turn.

The expanding brain

We know that modern humans have the largest brain size relative to body size, and also the largest neocortex and prefrontal cortex, of all animals (chapter 4). From the appearance of the first hominids, the brain size began to increase steadily, relative to the body weight, which was increasing also. Beginning at around 3.5 million years ago, there was a steady and accelerating increase (an exponential increase) in the size of the hominid brain, relative to body size, largely due to the increasing size of the neocortex, as discussed in chapter 4. During the past 2 million years of evolution of the line *Homo*, brain size doubled.

The steady increase in brain size was interrupted at around 1.5 to 2 million years ago by a 'bump' in the exponential curve caused by a somewhat more sudden

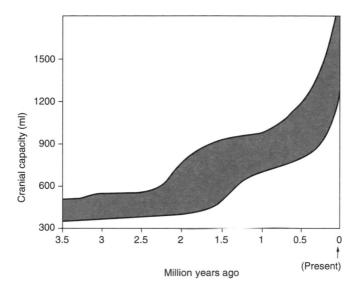

Fig. 5.3 The capacity of the cranium of the hominids increased over evolutionary time. This broad curve encompasses the data from the different specimens of fossil hominids (as in Fig. 5.2). Note that an increase in cranial capacity occurred around 1.5 million to 2 million years ago, followed by a hiatus until about 0.5 million years ago when cranium size began to increase rapidly. For more detail see Noble and Davidson, 1996.

increase in brain size (Fig. 5.3). It was at this time that the climate changed dramatically, and it continued to fluctuate considerably and over relatively short periods of time. Forests were lost over just several decades only to return again just as fast. William Calvin, a neurophysiologist at the Washington School of Medicine, USA, has hypothesised that these swings in climate may have caused the sudden increase in the size of the *Homo habilis* brain by promoting the accumulation of mental abilities that would permit flexibility of behaviour.

Such flexibility would be necessary for survival in the changing climatic conditions. An increased capacity of the brain would have made it ready for any new life style that might have been demanded by the changing climate. There is, however, no direct evidence to support this speculation, interesting though it is.

It is always taken as fact that increasing brain size means increasing cognitive complexity or intelligence. In a general sense, at least within one line of evolution, this may be somewhat true, but we must remember that species on divergent branches of evolution may use quite different organisations of neurons to solve the same problems of behaviour, and size is not always the issue. With their relatively small brains, birds can function at cognitive levels equivalent to those of primates (chapters 3 and 4).

Palaeoanthropologists, who study the evolution of humans by examining fossils, can only obtain information about the shape and size of bones. They can estimate brain size from the size of the cranium and body size from the skeleton, and it may be reasonably accurate, but they can only guess at the level of intelligence that an extinct brain might have had. The basic assumption of this kind of research is that brain size is directly and immutably related to intelligence. This assumption may be to some extent correct provided that one keeps within one undiverging line of evolution, but we will never know because intelligence is something that only a living animal can tell us, not brain size. This is the paucity of the hominid fossil record and it is on the unknowns and the cracks in the evidence that, all too often, our human-centred views are founded.

Standing on our hind limbs

Did brain size begin to increase as a result of hominids adopting an upright posture or was it the other way around?

Nonhuman primates move about by using all four limbs, either to move on the ground (e.g. baboons) or to swing in the trees and land on branches (e.g. many macaque monkeys and howler monkeys). In the case of our nearest relatives, the great apes, orang-utans use all four limbs in almost equal amounts as they move through the canopy of the rainforest. The hip joint of the orang-utan allows the legs to be moved more like arms, while the feet can cling more like hands. Chimpanzees and gorillas use four limbs likewise when they are moving through the trees but, when moving on the ground, they usually support themselves by using both their feet and the knuckles of their hands. They are referred to as 'knuckle walkers'. All of the apes are able to walk bipedally on the ground but they do not do so habitually, as we do. Also, when apes walk bipedally their gait is more laboured than ours because they cannot extend the knee joint to make a straight leg for stepping out and their feet have to be placed widely apart.

All hominids, except perhaps some of the earliest australopithecines, were bipedal. This can be deduced by the structure of their feet, hips and pelvic bones and of the joint between the bones of the neck and the back of the skull, as the head has to be held at a different angle when the body is in a bipedal versus a quadrupedal stance. In fact, 3.6 million years ago in the place we now call Tanzania, three hominids walked through some fine ash from a volcanic eruption. Their footprints were soon hardened by sun and rain and covered by more ash. In time, the footprints became fossilised. These ancient footprints, discovered two decades ago, showed that these early humans walked bipedally, a small one walking alongside a larger one, possibly parent and child hand-in-hand, and another following in the footsteps of the larger one. Judging by the size of the crania of skulls of about the same age as the footprints, these bipedal australopithecines would have had a brain size about the same as that of apes. Therefore, bipedalism might have preceded the increase in brain size that was to occur in hominids.

Various explanations have been proffered to explain why bipedalism evolved. I have mentioned already the one about more efficient movement over grasslands. The upright stance would also have allowed better detection of predators on the ground in long grass. It should be noted that some quadrupedal species, such as meerkats and vervet monkeys, adopt a bipedal stance when they are looking for ground predators. Although for an animal running at a fast speed bipedal locomotion is less efficient than quadrupedal, being bipedal may have enhanced the hominids' stamina for tracking prey at slower speeds; or, if early hominids were still vegetarian as the structure of their teeth suggests, they may have used their stamina to cover larger distances in search of plant foods or water. In addition, adoption of the upright posture would have freed the hands for carrying things and for throwing them. Thus the hominids could carry weapons for hunting, babies and vegetables or fruit gathered at a distance from the place where they were to be eaten. Bipedalism would also have freed the hands for using tools, although the earliest stone tools appear to have been used well after bipedalism evolved. It is, of course, possible that tools made of less durable material could have been used by hominids well prior to this time, as will be discussed later.

According to Dean Falk of the State University of New York, USA, bipedalism may have evolved for heat control. By living in open savannah, without the shelter of trees, the early hominids were exposed to the hot midday sun and, according to Falk, there may have been an evolutionary advantage gained by standing up away from the hot reflective substrate and at an angle that reduced the surface area of the body exposed to the direct rays of the hot noonday sun.

Standing upright had certain consequences for the brain, which requires a good supply of blood. There was a problem in getting blood to a head held upright high above the heart and also in getting the blood back to the heart without overloading the main vein involved, the jugular vein. Thus, along with becoming bipedal came certain necessary changes in the arrangement of the blood vessels and blood-carrying sinuses. The human skull became covered on its outside and inside surfaces with a complex web of communicating veins. This rearrangement of the blood vessels of the cranium could also serve to cool the brain. The brain requires a considerable supply of energy in order to function, and this creates internal heat (metabolic heat). The heat from the brain, therefore, needs to be dissipated, and the blood system that evolved along with bipedalism could be used to do just that. The network of veins could act like the radiator of a car to prevent overheating. Thus, Falk has argued that the change in the blood supply to the brain may have removed a major barrier for its expansion in size. With its new cooling device the brain could grow larger and so generate more heat.

Thus, according to Falk's hypothesis, the change in the vascular system of the brain may have evolved firstly to overcome the gravitational problem of supplying blood to the brain when australopithecines became bipedal, and then could have been used to cool the brain, in turn allowing the brain to expand. However, other species have evolved efficient ways of cooling the brain in hot climates (e.g. the nasal cavity of Nubian goats, and of camels and donkeys, acts as a recycling cooling device for the brain, and the ears of the elephant act likewise) and yet they have not shown any particular expansion of the brain along with this. Brain cooling cannot be the only factor that led to expansion of brain size.

There might be no single explanation for the evolution of bipedalism, and its consequences may not have been limited to a change in the vascular system of the brain. Bipedalism also freed the forelimbs and hands from their previous role of supporting the body. Hence, both hands could be used for carrying, for manipulation of objects, for tool use and for communication. As discussed in chapter 4, according to the hypothesis of Finlay and Darlington this newly acquired use of the hands may, itself, have been the driving force for expansion of the neocortex. Of course, this expansion may have been facilitated by the rearrangement of the cranial blood vessels that had already occurred.

Did bipedalism also lead to the right-handedness that predominates in modern *Homo sapiens*?

Handedness

Handedness is often cited by anthropologists and psychologists as one of the unique features of *Homo sapiens* that might reflect our superior place in evolution and, therefore, our consciousness. Humans are predominantly right-handed. Most of us use the right hand preferentially for manipulating objects, for writing and other acts that require fine movements. Most of us also use the right hand for hammering and throwing but, in fact, the degree of right-handedness in the human population is not as consistent or quite as strong as we usually think. The hand preferences of individuals vary quite considerably on different tasks. Few of us use the right hand absolutely consistently for all tasks. For example, a person may have a strong right-hand preference for writing but use the left hand for hammering or throwing and so on. Despite the fact that it is often claimed that humans are about 90 per cent right-handed, and this is true for writing, the handedness of the human population seems to be nowhere near as strong when a wider range of activities with the hands is assessed.

Recently, Linda Marchant, of the University of Miami, USA, and her colleagues William McGrew and Irenaüs Eibl-Eibesfeldt have used archival films to assess hand preferences in three traditional societies: the G/wi San of Botswana, the Himba people of Namibia and the Yanomamö of Venezuela. They scored hand use in a wide range of activities involving the hands using the techniques developed by ethologists to score the behaviour of animals accurately. The results showed the expected right-handedness but it was not as strong as the right-handedness that we associate with modern human cultures. Since the traditional people studied do not read and write, their weaker right-handedness might be due to not performing the activity of writing. To put it the other way around, in literate cultures the use of the right hand for writing may enhance right-handedness in other tasks as well as writing. However, it must be recognised that most of the data for literate cultures have relied on people filling in questionnaires about their hand preferences and this can give somewhat unreliable results. No one has yet scored hand use in literate human cultures by applying the same ethological techniques that Marchant, McGrew and Eibl-Eibesfeldt used for the traditional cultures. Were that to be done, the same weaker degree of right-handedness might be found in literate cultures also.

There was one form of hand use for which the G/wi San, Himba and Yanomamö people did show marked right-handedness, and that was tool using. They gripped tools that required fine manipulation with the right hand. Right-handedness appears to have been associated with tool using from the earliest time at which stone tools were used by humans. Nicholas Toth looked at the way in which the fracture patterns occurred on stone flakes made by early humans (*Homo habilis* and *Homo erectus*). The flakes were made in the manufacture of stone axes or flints for cutting. Toth concluded that the stone struck to produce flints must have been held in the left hand while it was struck from above by a stone held in the right hand. The fracture patterns fitted together in such a way that each strike would have produced a new flake as the rock held in the left hand was rotated clockwise relative to the blows with the hammer held in the right hand. Although Toth's conclusion has been contested on the grounds that the striking action might have been from below rather than from above, and therefore the opposite hand might have been used, I am most interested here in the conclusion that he reached, as follows: ' . . . early hominid tool-making populations were preferentially right-handed, a trait characteristic of modern humans but no other species. This argues for the development of a profound lateralisation of the hominid brain by 1.9 to 1.4 million years ago' (Toth, 1985, p. 611).

Handedness reflects specialisation of the hemispheres

(each hemisphere controls the hand on the opposite side of the body) and it is an aspect of brain lateralisation, discussed in chapter 4. Thus, Toth made an association between brain lateralisation, handedness and tool use. He suggested that handedness may have evolved in humans due to selective pressures to make tools and to use them.

Then another link was added to the brain lateralisation–handedness–tool-using chain of associations, and that was language. It was implied that consciousness is also associated with these characteristics.

As mentioned in chapter 4, language and speech are functions of the left hemisphere. Hence, the left hemisphere is specialised for controlling the right hand, and tool use by that hand and for communication using language. Some anthropologists have postulated that communication by gestures preceded the evolution of speech and thus right-handedness and tool use preceded language and led to specialisation of the left hemisphere for language. Others have suggested that designation of the left hemisphere for language came first and right-handedness followed. Yet others have gone as far as to speculate that language and the manufacture of tools may use very similar cognitive processes.

While there may, indeed, be similar or associated brain mechanisms for handedness, tool use and language, it is now clear that they did not evolve together. Handedness evolved very early in animals. Even toads have handedness, or perhaps it should be called pawedness. Recent experiments in the laboratory of Angelo Bisazza and Giorgio Vallortigara at the University of Udine, Italy, and by Andrew Robins in my laboratory at the University of New England, Australia, have shown that some species of toads prefer to use the right paw to wipe a small piece of paper from the snout or to push and pivot themselves to the surface of water when they have been turned over and submersed. Admittedly, the percentage of toads preferring to use the right paw is less than the percentage of humans that are right-handed, and some toads have no preference, but the bias towards right-pawedness is significant. This result suggests that forelimb preferences might have been as ancient as the first animals that moved out of water on to land. In fact, limb preferences might even have evolved amongst the fishes, before amphibians (e.g. toads). Some species of fish show biases to swim in a particular direction of turning. For example, when these fish see a predator almost all of them turn in the same direction, either leftwards or rightwards depending on the species. Also, Michael Fine and his colleagues in Virginia, USA, have reported that channel catfish prefer to rub the right fin against the pectoral spine in order to produce a pulsating sound. This behaviour is equivalent to handedness.

Pawedness occurs in other species too. One study has reported that dogs prefer to use the right paw to wipe away sticking tape from their eyes, not a very pleasant experiment. There are more studies of paw preferences in cats, and they indicate that cats prefer to use the left paw to reach for and grab food or moving objects. Some species of birds have foot preferences. Most parrots and cockatoos, for example, prefer to hold food in the left foot. In fact, I have found that sulphur-crested cockatoos are so strongly left-footed for holding food that I have yet to see a right-footed one, although I am sure that some right-footed ones do exist. The footedness of some species of birds is as strong as or even stronger than the handedness of humans.

Primates, too, have handedness, despite earlier claims that they did not. It used to be thought that, in any species of primate, some individuals have a left-hand preference and others a right-hand preference and that these balance each other out so that there is no overall bias, or handedness, in the population. That is, primates were thought to have nothing akin to the right-handedness of humans. Such a 50:50 distribution of hand preferences is, in fact, characteristic of rats and mice, but not of primates. As Jeannette Ward of the University of Memphis, USA, has shown, among the lower primates (the ones that evolved first, lemurs and bushbabies), left-handedness predominates for picking up and manipulating food objects. According to the hypothesis of Peter MacNeilage, some of the monkeys are left-handed, whereas the apes have a tendency to be right-handed. He hypothesised that the right hand of primates is the strong hand and that it is used for holding onto branches while the left hand is used for reaching for food and taking it to the mouth to eat, as in the lower primates. According to this hypothesis, once primates became a little more bipedal, as in the case of the apes, the right hand was freed from having to support the body and could be used to manipulate objects. The left hand is better at grabbing moving objects and the right is better for manipulation. This seems to be true for many species. Which hand gets used in a particular situation depends on whether accurate grabbing or fine manipulation is required.

There is still debate about handedness in apes. For example, chimpanzees raised in captivity appear to be right-handed, like humans, whereas wild ones may not have a population bias for use of one hand over the other, at least according to the observations of William McGrew and Linda Marchant. It seems that hand preferences might be modified by the amount of practice at climbing, contact with humans and the nature of the task being performed by the hands. To illustrate the last point, Gisela Kaplan and I have studied hand preferences in orang-utans in Sabah, East Malaysia, and found that, although there was no bias for all orang-utans to use the same hand to hold and manipulate food, there was a very strong population bias for them to use the left hand to manipulate parts of their face, for example, to clean the teeth, nose or ears. Humans, apparently, show the same left-hand preference to touch the face. This finding of left-handedness in orang-utans is important because it demonstrates that orang-utans have a lateralised brain like humans and that the strength of this handedness is equivalent to that of humans. It also shows that handedness is not a unitary characteristic that appears in all tasks but, rather, it may be present for one type of hand using and not another.

In chapter 4, I discussed some of the now quite exhaustive evidence for lateralisation of the brain in animals. All of this information on animals has been accumulating over the last two decades but, rather surprisingly, little of it appears to have been taken into account by anthropologists. By the time that Toth stated that right-handedness was a trait characteristic of modern humans and no other species (quoted previously), lateralisation of the brain and footedness in birds had been well documented. From their human-centred perspective, anthropologists are, of course, not interested in birds, and it was not until 1987 that MacNeilage and his coauthors published their paper on handedness in primates. However, even that and the flurry of reports on handedness in primates that followed were ignored by Richard Leakey in his book *The Origins of Humankind* published in 1994. He still claimed handedness, language and tool use are unique to humans, as seen by the following quotation:

> Although individual apes are preferentially right- or left-handed, there is no population preference; modern humans are unique in this respect. Toth's discovery gives us an important evolutionary insight: some 2 million years ago, the brain of *Homo* was already becoming truly human, in the way that we know ourselves to be. (Richard Leakey, 1994, p. 41)

These words demonstrate how one field of science can ignore another and how reluctantly favourite theories are discarded. Even within one field there can be blind spots: the authors, mentioned previously, who wrote about the lateralised fin use in catfish stated incorrectly that primates and other mammals lack handedness, even though it was well known at the time the paper was written.

Other people have recognised that lateralisation of the brain and handedness are not unique to humans, but have attempted to keep alive the theory linking human evolution to language and lateralisation by claiming that lateralisation in humans is greater than in animals. This is not so. We already know that chickens are just as strongly lateralised as we are, and that they have lateralisation of just as many functions as we do. Footedness in some species of parrots is also as strong as handedness in humans.

At the time that Michael Corballis wrote his book *The Lopsided Ape* (published in 1991), he may have been correct in saying that handedness in nonhuman primates is weaker than the handedness of humans, but the handedness of orang-utans for touching the face is, in fact, as strong as the handedness of humans. Therefore, I do not agree with the following statement: 'The critical events that shaped our handedness must therefore have taken place since the time that the split between humans and chimpanzees occurred' (Michael Corballis, 1991, p. 99).

Also, I think that rather too much emphasis is placed on handedness. It is only one manifestation of brain lateralisation. There are many other forms of lateralisation and these too are not exclusive to humans either in kind or degree. Over a wide number of species (reptiles, birds, rodents and primates including humans), the left hemisphere is specialised to process and make the vocalisations typical of the species and the right hemisphere is used to assess spatial positions of objects and to control emotional behaviour. Thus, being handed or having a lateralised brain are not unique characteristics of humans—they are not intimately associated with language or the kind of consciousness which is present in humans.

Tool using

Tool using may require a special aspect of handedness. In humans, tool using that requires fine control with the fingers in what is referred to as a precision grip (as opposed to a power grip) is predominantly carried out by the right hand. I have mentioned already that Toth presented some evidence that early humans had made flints by holding a stone hammer in the right hand and striking it against the stone from which the flints would come, held (almost certainly in a power grip) by the left hand. In addition, most prehistoric stone axes are made for right-hand use. Perhaps the same hand was preferred because axes and other tools could then be shared, or perhaps it was easier to learn how to make a tool if it was an exact replica of the prototype rather than the mirror image. Both hypotheses have been put forward by archaeologists.

The extensive evidence for handedness in primates and many other animals shows that handedness evolved well before tool using. Despite this, it does remain possible that tool using enhanced right-handedness for the reasons that the archaeologists have suggested. The G/wi San, Himba and Yanomamö people were most strongly right-handed when they were using tools and this is likely to be true for other human cultures, given that we construct scissors, saws, pots for pouring and most other tools so that they can only be used effectively when held by the right hand.

Some researchers in this field have implied that right-handedness in tool using is unique to humans, and they cite evidence that they have collected for hand preferences in wild chimpanzees fishing for termites (described in chapter 3). William McGrew and Linda Marchant scored the hand in which the chimpanzees held the piece of twig when they were inserting it into the termite mound. Of the fifteen individuals that they scored, six had left-hand preferences for holding the twig, five had right-hand preferences and four had no hand preference. Thus, there was nothing equivalent to the right-handedness of humans in tool using. Similar use of a probing tool has been scored in a small group of captive South American capuchin monkeys (*Cebus apella*) and most of these used the left hand to hold the probe.

Wild chimpanzees also use tools to crack open nuts, as explained in chapter 3, and they usually hold the hammer in the left hand. Yukimaru Sugiyama and colleagues from Kyoto University, Japan, found that adult chimpanzees at

Bossou in West Africa held the hammer stone by preference in the left hand, whereas juveniles in the same group used either the left or right hand and thus had no group preference (i.e. there was no handedness in the juveniles). This result suggests that the consistent left-hand preference in adults is learnt and it may be related to more success in cracking the nut when the hammer is held in the left hand. Like chimpanzees, capuchins use the left hand to hammer nuts, but they use the right hand to use leaves as a sponge. These contradictory data on hand preferences indicate that we need to collect a lot more information on hand preferences in tool using by both wild and captive primates, and to take the age of the subjects into account, before we can draw any conclusions about the uniqueness, or otherwise, of human right-handedness in tool using.

Also, one can only assume that humans would use the right hand to insert a twig into a hole to fish for termites. This has, however, never been tested. As discussed in chapter 4, in a very wide range of species the right hemisphere (which controls the left hand) is specialised to perform tasks that rely on spatial information. As a consequence, right-handed humans are quicker and more accurate at reaching out to grab a moving object with the left hand. On this basis, we might predict that humans, as well as chimpanzees, would be more accurate at inserting the twig into the hole, using spatial information, when they use the left hand rather than the right. The important thing to measure might be accuracy, rather than which hand is used more often. Unfortunately, this has not been done for humans or chimpanzees. Also, it would be important to do the scoring at the beginning stages of performing the task because, with practice, the right hand might become as accurate as the left.

The right hand of both chimpanzees and humans might be better at manipulating the twig using fine finger movements (controlled by the left hemisphere) and, of course,
bility could be useful in certain aspects of termite
ch as turning or moving the twig around when

it is in the hole or bringing the termite-laden twig to the mouth. Thus, there may be a trade-off between the hands to be used in this task: better ability to insert the twig with the left hand might be balanced against better manipulation of it with the right hand. The choice to use the left or right hand may depend on the species or on past experience and have nothing to do with being an animal as opposed to a human, as some have claimed. The right-handedness of humans and, indeed, of capuchin monkeys in certain tool-using tasks may be more to do with what information they are using to process the task than something unique about either species. In nut cracking, the left hand may be used because spatial aspects are important for striking the nut, but the nut is positioned on the anvil ready for the strike by the right hand, the one specialised for fine manipulation.

Many examples of tool using by various species of animals were described in chapter 3. Tool using is not exclusive to humans but, of course, we use a greater variety of tools in more complex ways. This may be a reflection of our more highly evolved brains and it may also have been one of a number of factors that, somehow, drove the evolution of a more complex and larger brain.

Homo habilis was making stone tools around 2 million years ago. Sharp flakes and the stones from which they were chipped have been found. The flakes appear to have been used to cut plant material or meat, or to manufacture other tools, such as digging sticks. If this was the first appearance of tool use by hominids, it coincided with the rather sudden increase in brain size mentioned earlier, but it is possible that earlier hominids were using tools made of less durable materials. Thus, *Homo habilis* might have been the first hominid to use *stone* tools, but not the first hominid to use tools as such. There may, of course, be something special about using stone tools but I suspect that this would be region specific. In certain parts of the world it may be most important, and only possible, to use one kind of tool and in other regions another kind of tool.

Thus, *Homo erectus* in the Nihewan Basin of China developed simple stone tools, whereas *Homo erectus* in Southeast Asia may have specialised in tools manufactured from bamboo, which leaves no archaeological trace. There is nothing to say which kind of tool material, stone or something less durable, requires a higher cognitive capacity, although the weight of thinking in archaeology is on the side of stone tools.

Nor, in my opinion, does (stone) tool making signal the appearance of consciousness. Certainly, to make a tool requires planning ahead, and this depends on at least one aspect of consciousness, but planning ahead can also be manifested by other behaviours that do not leave archaeological records. To link the appearance of intentionality (planning ahead and behaving with a purpose in mind) to the appearance of stone tools is, I believe, mistakenly based on what manufactured objects leave an archaeological record. Moreover, the same planning ahead is required to make a wooden tool as to make a stone one, and to make one tool as to make many. In chapter 3 the ability of capuchin monkeys to make tools from bamboo was discussed and, together with the now extensive examples of tool manufacture and use by apes, this suggests that tools made of perishable materials were being used well before humans evolved. The step made by *Homo habilis* to make simple, asymmetrical stone tools may not have been particularly unusual. It was not until much later that *Homo erectus* began to manufacture symmetrical tool forms that were often fashioned around fossilised shells in the rock. These decorated stone handaxes might have signalled the first appearance of artistic representation in hominids or, as some argue, they might have been ornamented merely by chance due to the possible ways of fracturing rocks with fossils in them. Scepticism and debates abound, but they must now take into account the evidence that the ability to form mental representations evolved well before humans (see chapter 3). Moreover, the kind of planning ahead that must characterise the making of tools also evolved well before humans, as shown by tool manufacture in wild chimpanzees and orang-utans, in particular.

Tool using does not appear to have been, in itself, particularly associated with the expansion of the hominid brain. From its appearance around 2 million years ago it developed extremely slowly, with the development of a tool kit at 1.5 million years ago, and there was no other major advance until around 300 000 years ago with the development of carved spears of beautiful shapes. Tool using does not appear to be either a reflection of or a driving force for the enlargement of the human brain, although it might have had a stronger relationship to a subregion of the brain.

Language

Although some forms of communication in animals share some aspects of human language, as far as we know no form of animal communication is as flexible, creative or complex as human language. We should recognise, however, that this opinion may have been reached because we know too little of any form of communication in animals. This remains an open possibility but not one that can be resolved here. We do know, however, that bird song has surprising similarities to human language in terms both of its development and complexity.

In many species, including frogs, birds, rodents, monkeys and humans, the left hemisphere is specialised for communication by vocalisations (chapter 4). One important characteristic of the brain concerned with the comprehension of language and the production of speech by humans is the greater involvement of areas in the left hemisphere compared with the right hemisphere. If a person has a stroke that causes damage to the left hemisphere, the inability to speak or to understand language may result depending on exactly which region(s) of the brain is (are) damaged. In most people, there are two major regions, called Broca's area and Wernicke's area (Fig. 5.4), in the left hemisphere that are concerned with speech and language. Broca's area is

153

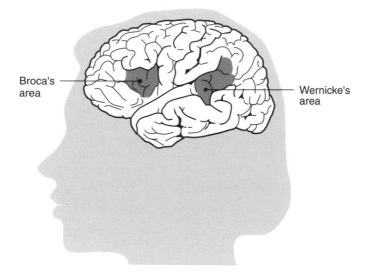

Broca's area

Wernicke's area

Fig. 5.4 A view of the left side of the human cortex showing the regions involved in speech and language, Broca's area and Wernicke's area.

involved in speech production and Wernicke's in the comprehension of language. The Sylvian fissure, which can be seen as a groove on the surface of the brain, is longer and positioned lower on the surface of the left hemisphere than it is on the right. This asymmetry reflects the presence of Broca's and Wernicke's areas in the left hemisphere only.

The Sylvian fissure is asymmetrical also in the other apes and some species of monkeys, which may suggest precursors to human language. As the left hemisphere is specialised for producing and processing the communication signals of a number of animal species, it would seem that the areas of the human brain involved in language and speech may have evolved from equivalent areas in animals. There are also other anatomical asymmetries common to humans, other apes and monkeys. In apes and Old World monkeys, as in humans, the left occipital lobe of the cortex (at the back) is larger than the right and the right frontal lobe is larger than the left.

It is possible to identify the presence of Broca's area in a brain by the arrangement of the overlying grooves (sulci) on the surface of the brain. These sulci leave impressions on the inside surface of the skull. Thus, by examining the skulls of the extinct early hominids it should be possible to determine when a distinct Broca's area might have evolved and to deduce from this when language might have first appeared. Dean Falk, mentioned earlier, found evidence that Broca's area was present in *Homo habilis* 2 million years ago. Falk made cranial endocasts of skulls of primates and hominids. This involved filling the inside cavity of the skull with latex rubber and removing it after it had set. The procedure gives a model of the brain that was in the skull, and the sulci on its surface can be seen, although not always with great clarity.

Using this method on a skull of *Homo habilis* known as KNM ER 1470, collected from Kenya and thought to be around 2 million years old, Falk was able to see evidence of Broca's area. Of course, presence of a brain region is not conclusive evidence that it was, in fact, used for speech. In chapter 4, I mentioned that neurons that process auditory signals (sounds) can grow into and take over the main region of the cortex usually devoted to processing visual information, if blindness occurs from birth. Thus, the functions of particular regions of the brain are rather flexible, and they can change if an abnormality occurs during early development of the brain. Therefore, Broca's area, or what looks like it from the rather crude impression made on the skull, could, perhaps, have been used for some function other than speech. This is what William Noble and Iain Davidson of the University of New England, Australia, think. They believe that language appeared much more recently than 2 million years ago. In their opinion, language was only starting to make its appearance as recently as 100 000 to 70 000 years ago, when humans might have been building boats to make planned migrations (e.g. from Asia to Australia), and that it was definitely present only as recently as some 32 000 years ago when humans were making symbolic representations in bone and stone (see later).

These dates for the origin of language can be only reasoned guesses. It has to be recognised that the existence of Broca's area in *Homo habilis* makes it distinctly possible that humans were, in fact, using language 2 million years ago. Maybe it was a rudimentary form of language but, if so, why was Broca's area so well developed that it left a recognisable impression on the skull?

There is a stronger piece of evidence against the hypothesis that the Broca's area present in brains 2 million years ago was used for some function other than speech. Although, as I have said, the *developing* brain has remarkable flexibility, allowing one region to take on the function of another if some abnormality occurs (e.g. blindness), this is not so for the *evolving* brain. Evolution and development are often confused. Comparative neuroanatomists, who compare the structures and functions of the brain in different species, are always impressed by the conservation of structure and function across species. Of course, there are differences between species but evolutionary connections can be made. For example, once a particular distinct region of the brain has evolved to have a particular visual function (e.g. for detecting moving visual stimuli and locating their position), it tends to retain that function as evolution proceeds and new species form. During the course of evolution the function may be modified, and perhaps improved, but basically the designation of that region to perform a particular function is retained. Only if an abnormality occurs during the development of the brain might the function of a particular region be switched to something other than the role that it has been assigned by evolution. What does this mean for Broca's area in *Homo habilis*? I would say that, if it was not designated for speech as we know it in modern humans, it was used for something very close to it.

The ability to speak requires not only the appropriate regions of the brain but also the correct apparatus to produce the sounds. It must be possible to move the tongue into the correct positions, and the larynx (voice box) must be in the right place. In apes the larynx is positioned higher up in the neck and they cannot make speech sounds. The larynx had to descend in the neck before hominids could make speech sounds. There is controversy about when that occurred. Some say that it happened as recently as 30 000 years ago and others that it happened much earlier, in *Homo habilis*. Yet others have reasoned that the tongue is more important for speech than the larynx and that *Homo erectus* had the tongue muscles attached to the jaw bone in a manner that would have permitted speech. There is no solution to this controversy, but it should be noted that birds that mimic human speech produce speech sounds with a vocal apparatus entirely different from that of humans. Sea lions can also produce speech-like sounds. In other words, there may be ways around making the vocal apparatus work to produce speech sounds even if it is not easy or perfect and as long as the brain has developed the capacity to control speech.

It remains possible that *Homo habilis* of 2 million years ago might have been both speaking and making tools. Indeed, he or she might have been speaking about making tools. It may be pure coincidence, but the regions of the brain that control the mouth movements of speech are located right next to those that control the hands. Some people have argued that speech and hand use evolved 'hand-in-hand' with each other. Communication by means of gestures might also explain this association. Even in modern humans of today, hand gestures occur with speech and they follow the same rhythm as speech. In *Homo sapiens* speech and fine control of the hands are closely related to each other, but that does not mean they evolved at the same time.

157

Mental representations and art

The ability of both animals and humans to form mental representations was discussed in chapter 3. In humans, mental representations may be expressed in art forms. The earliest symbolic art forms of humans that have been unearthed from European sites date back to a mere 32 000 to 40 000 years ago, although recent finds of rock art in Australia by Richard Fullagar of the Australian Museum and colleagues may set this date back to about twice as many years ago. The latter finding is a matter of controversy in archaeological circles, depending on the method used to date the samples. Irrespective of this debate, the expression of art in durable media is a relatively recent development of the human species. On the grounds that language is a symbolic communication system, Noble and Davidson reason that the origin of language is also recent and that it coincided with the appearance of these symbolic art forms. There are at least two pieces of evidence against this hypothesis. First, as discussed already, Broca's area of the brain was present well before the appearance of the symbolic art forms. Secondly, prior to this time, there may have been less durable symbolic art forms, such as weaving or carving of soft wood, in which humans expressed their internal representations. These would not have survived to be discovered by the archaeologists of today.

Archaeologists of Western cultures rely on art forms that are expressed in materials that persist, such as carvings in bone, ivory or rock or paintings on rock, whereas, even today, many art forms are expressed in transient media. The ancient Japanese art form of ikebana (flower arrangement) is not less aesthetic or symbolic because it is transient. In fact, its very transience is part of its symbolism. We cannot know whether our human ancestors used such art forms, as they leave no tangible trace. Likewise dance and song may have been used as symbolic expression well prior to the making of sculptures and paintings. Perhaps they were a logical progression from displays in animals. Might not we think of the ritualised 'dancing' of animals (called displays) as symbolic communication? Of course, one could argue that the displays of animals are not intentional forms of communication, and that is likely to be correct for some species (e.g. the honey dance of bees) but perhaps not for other species. Where one might draw the line on intentionality in displays is more a matter of opinion than substantive fact, and that would be true for the displays of hominids as well as animals.

The ability of animals to form mental representations is not, as some have claimed, unique to humans, although it might be that humans can hold mental representations for longer and do more with them (e.g. compare one with another) than can animals. This is not yet known.

There is some evidence that the frontal lobes (which contain the prefrontal cortex, mentioned in chapter 4, as well as Broca's area) of the mammalian cortex are used for forming mental representations and for keeping them in mind to guide behaviour. Of course, this is not likely to be the only function of the frontal lobes, but it could be a most important one for the kind of consciousness that allows planning ahead and dealing with symbols. It is well known that people who have had frontal lobotomies (severance of the frontal lobes from the rest of the brain, used as a highly dubious treatment for depression) experience disturbances of attention and impaired ability to plan ahead. They may also show a 'flattening' of emotional reactions and changed, sometimes inappropriate, social interactions. At this time, it would be misleading to say that the functions of the frontal lobes are known with any degree of certainty but the indications are sufficient for making speculations.

Compared with those of other primates, the frontal lobes of humans are very large. Falk has used the cranial endocasts of hominid skulls to look at the size of the frontal lobes relative to the rest of the brain, and she has concluded that there was a particularly dramatic enlargement of the frontal lobes in the evolution of the *Homo* line to modern humans (Fig. 5.5). The convolutions of the frontal lobes

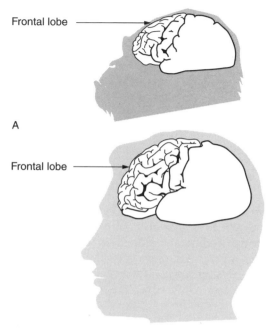

Frontal lobe

A

Frontal lobe

B

Fig. 5.5 The frontal lobes of a chimpanzee (A) and a human (B). Note that the frontal lobe of the human is larger in proportion to the rest of the cortex than is the frontal lobe of the chimpanzee

increased as the size of the frontal lobes increased and they were detectable from their impressions on the skull. Maybe this anatomical change to the brain reflects the evolution of the human abilities of mental representation and consciousness but we are, once again, reminded that 'bigger is better' is an assumption and that we cannot prove this with fossil material.

Nor can the ability to make mental representations be exclusively tied to the frontal lobes. In birds another part of the brain must be used to form visual representations as, for example, can be formed by young chicks. This ability of young chicks might be vestigial compared with human abilities to form mental representations but the fact that it occurs in the absence of any frontal lobes is of interest. Also, until the recent experiments that I discussed in chapter 3, no one would have credited birds, particularly such young ones, with the ability to form mental representations. These results for animals might force us to think differently about humans.

Certainly, there might have been a so-called 'creative explosion' that occurred in humans 30 000 to 40 000 years ago or, at least, there was a cultural shift to express mental representations in nonperishable forms. Although this must tell us something important about humans and their culture, I suggest that it is not an event on which to pin the appearance of either language or consciousness. This time might, however, have been important for the flowering of language, culture and consciousness.

Society, superstitions and the hominid mind

In chapter 2, I discussed the topics of mind-reading and deception and how they can be used to advantage in social situations. These behaviours are not unique to humans. There is some evidence that chimpanzees can attribute mind states to others (to members of their own species and to humans too) and there is also evidence suggesting that tactical deception may occur in many animal species. *Homo sapiens* may, of course, make greater use of these tactics than does any other species. According to Alison Jolly and Nicholas Humphrey (discussed in chapter 3), higher intelligence evolved with increasing social complexity. The evolution of consciousness may be associated with this. Human societies are seen as the most socially complex of all, and hence we consider ourselves to be both more intelligent and more conscious than other species. Many anthropologists believe that consciousness must have

161

developed somewhere in the *Homo* line of evolution, but how convincing is the evidence for this?

One may speculate that consciousness emerged when the brain took a leap forward in size and the neocortex developed a sufficient degree of complexity but this is, indeed, purely speculation. Consciousness can only be measured either by looking at behaviour or by listening to what another person says. As we have seen, we can look at the behaviour of living species and try to assess whether they have consciousness, difficult though that might be. We do not have this kind of access to the behaviour of the now extinct hominids. By examining the traces of their life style and the relics that they have left behind, we can make some deductions about their level of skills and, with reservations, we can deduce something about their intelligence. Can we tell anything about their consciousness?

Anthropologists have asked when it was that hominids started to think about the future. Burying bodies could be taken as an indication of consciousness of something occurring in the future. Superstitions that are part of the ritual of burying involve thinking about events or images in another time and place. They are the manifestation of a certain kind of consciousness. However, although superstition or religion is a major aspect of burial in all modern humans, this may not have been the case when burial first became a practice of the hominids. Burial also serves to cover over decaying matter and thus may have represented a straightforward biological advantage that later became associated with superstitions. As such, burying may, in the first place, have been very little different from the burying of bones by dogs, acorns by squirrels or seed by the storing birds that I discussed in chapter 3. Also, worker bees remove the bodies of dead drones from the hive. Evidence that burial occurred does not, unfortunately, tell us that humans did, in fact, worry about the future, although some anthropologists have assumed that it does.

The first burials have been attributed to the Neanderthals, which existed 0.12 million to 0.04 million years ago, as a side branch of the *Homo* line of evolution that did not lead directly to *Homo sapiens*. Some anthropologists contest this early date for burials, claiming that apparent burial may simply have happened by accident when, for example, a cave roof fell in on a sleeping Neanderthal. If so, burial appeared much more recently in the hominids.

Irrespective of when hominids began to carry out deliberate burials and whether burial tells us something about the appearance of superstition in hominids, I do not believe that it signals the beginning of being able to plan ahead. Nor do I believe that planning ahead appeared as late as making boats to migrate, as Noble and Davidson have said. In chapters 2 and 3, examples that may indicate planning ahead by animals were discussed. It would be unwise to pin the rise of consciousness to the archaeological indications of planning for the future.

What can we conclude?

To draw general conclusions from scattered information based on a number of assumptions is always risky but I believe that it can be said that tool using, language, culture, social complexity, high intelligence and consciousness all came together with the evolution of humans. Not one of these characteristics appeared for the first time in humans, despite the fact that this is often said to be the case. One could say that the evolution of humans was the drawing together of threads representing each one of these characteristics that appeared many times over in different forms in different species. If there is a discontinuity between *Homo sapiens* and other living species, it does not lie in the exclusive possession of any one of these traits. Other animals use tools but we use more of them and more complex ones. Other animals have complex communication systems that share aspects of human language. They may be less sophisticated than human language, although they are probably far more sophisticated than we presently understand. The kind of consciousness that *Homo sapiens* has may be special, but we are not likely to be alone as the only species that is aware of itself. Symbolic language might have extended the power of our minds and it must have enriched consciousness but, in my opinion, it did not mark the first appearance of consciousness.

There is a continuity of human speech with the brain structures that are used for vocalisations in animals. Lateralisation is as typical of animal species as of humans. Both stone and wooden tools were being used well before humans evolved and planning ahead is essential to the survival of many species. No single feature on its own makes us special.

CHAPTER SIX

FUTURE RESEARCH ON
ANIMAL MINDS

We have a long way to go before consciousness in animals has been fully established as a *scientific* fact, despite all of the indications of its existence that have been described in previous chapters of this book. In chapter 1, I said that lack of a unitary definition for consciousness should not inhibit research on the topic, but we should not forget that different researchers may be looking for different things. Attention could be focussed on research on one particular facet of consciousness, but it is difficult to choose what might be the best facet to look at first. There is also a danger inherent in a focussed approach and that is the risk of that single approach becoming the axiom for all further research on consciousness. Were that to happen, it would distort or stifle other approaches as, for example, has occurred with IQ testing and research on intelligence in humans. Performance on an IQ test (which gives a numerical result, called the Intelligence Quotient) is only one aspect of the much broader collection of attributes that were referred to as intelligence, but IQ has dogged the field of research on intelligence in humans for decades. With this in mind, I think that research on cognition and consciousness in animals should proceed along its many different directions but that it should take more account of several issues that I will outline in this chapter.

The present flowering of scientific investigation into consciousness in animals is coloured by our attitudes to animals. There is much at stake in the social realm: human societies have always relied strongly on either coexisting with animals or exploiting them. It is these attitudes that shape our approach to the science of animal cognition and consciousness. Scientists, it is said traditionally, must enter into research from an unbiased position and interpret their findings in the same manner. As Steven Rose and Hilary Rose made clear some years ago in their book *Science and Society*, scientists do not work in ivory towers shielded from the attitudes of society. We enter into any research on animals with a history of ideas about animals that have reached us through our culture in the wider society and from within scientific disciplines that prescribe certain attitudes to our research subjects. These attitudes are most evident in the investigation of consciousness in animals.

Attitudes and the case for or against consciousness

Scientists researching animal consciousness may hold opposing positions. Gabriel Horn of Cambridge University, UK, who researches memory formation in chicks, has said that chicks have memory systems very similar to those of humans. Writing in 1988 he said that, when an animal behaves in such a way as to satisfy the criteria for judging the state of consciousness in human beings, it seems logically capricious to argue that the animal is not conscious. He also stated that he suspected that the time will come when the view that humans alone are conscious will be regarded as being as ignorantly anthropocentric as the view that the sun revolves around the earth. This position is in contrast to that of psychologist Celia Heyes, University College London, UK, in her 1993 critique of the methods that have been used to study deception and attribution of mental states in animals. Heyes is of the opinion that, until there is definite proof that animals can attribute mental states and are not responding in simpler ways, the null hypothesis must continue to be that animals do not attribute mental states.

Thus, Heyes will hold the Descartian position that animals are assumed to lack the ability to attribute mental states (consciousness) until they are proven to be able to do otherwise. Although she says that she has no intention of stifling research on mental states in animals, she points out that it is, and will continue to be, extremely difficult to prove (beyond doubt) that animals have consciousness. In other words, it is going to be hard to convince her, as it will be many other scientists, that animals have consciousness. Compared with Horn's position, hers is a closed one. It is, most certainly, desirable to adhere to strict scientific rigour when gathering evidence, but to adopt an unswerving position against consciousness in animals until it is proven otherwise is a matter of opinion, not scientific rigour. Horn, Heyes and any other scientists may meet the same criteria of rigorous investigation irrespective of whether they begin from an open or closed position about the topic. In most circles of scientific enquiry, however, the closed position receives more kudos, much to the detriment of animals, as I will explore in chapter 7.

Much of the information on deception and 'mind-reading' in animals comes from anecdotal reports made by researchers studying wild animals in the field, a number of examples of which were given in chapter 2. Heyes rejects this evidence on the grounds that it is rarely possible to tell whether a given observation has been fortuitous or not. She is more predisposed to controlled experiments carried out in laboratory conditions, such as the work by Povinelli on chimpanzees, although she has objections to some details of his particular experimental methods. As I said in chapter 2, the chimpanzee is required to know the mind states of two humans, one who knows which of four cups has been baited with a morsel of food and another who does not. The 'knower' signals correct information about which cup has been baited to the chimpanzee and the 'guesser' signals cups at random. Chimpanzees were able to learn to follow the knower and thus had attributed a state of knowledge to that particular person. Although this result is quite convincing, Heyes has suggested that they might have merely learnt to respond to subtle cues given by certain movements or directions of eye gaze of the testers. In fact, Heyes pointed out that there were differences in the ways in which the 'knower' and 'guesser' moved and looked in the tests given to the chimpanzees as opposed to their movement and appearance in tests that Povinelli also gave to monkeys. This could have explained why Povinelli concluded that the monkeys were unable to attribute mental states whereas the chimpanzees were able to do so. It is important to draw attention to these possible influences on the results and I agree with Heyes that it is important to approach all scientific research logically and with controlled experimental procedures.

Unfortunately, however, tightly controlled experiments usually demand rather sterile and contrived testing environments that may counteract the expression of complex cognition and evidence of consciousness. From photographs of the testing apparatus used by Povinelli and colleagues, one can see that the chimpanzee is being tested in a rather sterile laboratory setting, much like the clinical environment of a hospital. Many readers will be familiar with the disconnected, dazed state of mind that one develops after a period of time in a hospital ward. It is known that humans perform differently on many tests when they are given in such an environment compared with their performance outside in the 'real' world. In fact, the mind state in a hospital environment is so different that patients that have been treated with a psychoactive drug (e.g. a major tranquilliser) in hospital may react quite differently to the same drug when they leave hospital. In fact, the differences in the physiological and psychological responses of the same patient in different social situations is so well known that Patricia and Jack Barchas of Stanford University, USA, have developed a separate field of study called 'Sociopharmacology' to investigate the effects of the environment on drug reactions.

Thus, although of interest for the very fact that experiments in the laboratory can be controlled in ways that studies of wild animals cannot, results obtained from captive animals should not be seen as limits to the species as a whole. A monkey tested in the laboratory may never show that it can attribute mental states to others, but that does not mean that other members of its species in the wild may not be able to do so.

Added to this, the experiments used to test for evidence of consciousness are often extremely contrived. In some of Povinelli's tests the guesser stood with a bucket over his head while the cups were baited. How often would anything like this occur in the chimpanzee's real world? Even the procedure of pointing to hidden food is unlikely to occur amongst wild chimpanzees. The fact that the chimpanzees tested like this displayed the ability to attribute mental states is, perhaps, a tribute to mental abilities far in excess of those being used in the task!

Differences between species in the way they react to the same testing situation is often ignored. It is common for monkeys of various species to be compared with chimpanzees by testing them all on the same task. Using such procedures many researchers have concluded that monkeys lack the ability to attribute mental states to others whereas chimpanzees can do so. If all species are given the same kind of test, there are bound to be those that have the ability that is being tested and those that are found to be wanting. As we have seen with the test of self-recognition in a mirror, species differences in sociability may influence the results of the test, and so too might differences in attention to the part of the body to which the spot of coloured dye is applied (see chapter 2). The original conclusion that apes could recognise themselves in the mirror, whereas monkeys could not, did not take these factors into account.

All too often, the results that have been obtained by testing a few chimpanzees are said to characterise 'the chimpanzee' in general, as a species. 'Chimpanzees' are said to be able to attribute mental states and to contemplate and solve problems, whereas 'monkeys', it has been claimed, cannot do either of these things. The very few individuals tested cannot represent the entire species, but even more astounding is the fact that although there are hundreds of species of monkeys they are so often referred to collectively as if they were one species. The different species of monkeys are adapted for different environments, have different social behaviour and different physiology, and must have very different 'intelligences' or mental states (chapter 3). In order to understand the mental processes in animals, these kinds of sweepingly inaccurate claims need to be set aside.

This does not mean that there are no characteristics that are shared by all, or most, members of a species, or that we will never be able to discover the mental abilities that are characteristic of a species. We already know many behaviours are typical of particular species. However, the path to concluding that a particular behaviour or performance ability is species typical must be trodden with caution. Just because a small group of monkeys of a single species does not, for example, exhibit the ability to attribute a state of mind or knowledge to another in one particular testing situation, it does not mean that all monkeys in all situations would behave likewise.

We should also remember that animals not only tested but also raised in laboratory conditions have all been 'institutionalised', and we know from humans that this existence tends to suppress at least some aspects of complex cognition. When the problem-solving or language abilities of nonhuman apes are compared with those of humans, no mention is ever made of the fact that in the majority of cases the nonhuman apes have been living in relatively impoverished laboratory or zoo environments, whereas the humans with whom they are compared have suffered no such deprivations. Of course, one could consider that the special language training that the apes received actually enriched their experience but, if it did so, it was in a particular framework, not in a general sense.

The chimpanzees that Beatrix and Allen Gardner (chapter 1) taught to use sign language were raised in an environment that was similar to that of human children and thus not impoverished, although the chimpanzees' situation was very different from being raised in the wild. Some scientists, however, have criticised the Gardners' original research on the grounds that the rearing conditions were not controlled rigorously. Here is a double bind. On the one hand, the rearing and testing conditions must be controlled completely or the complex cognitive abilities that animals display will not be believed. On the other hand, if the rearing and testing conditions are controlled completely, the environment becomes so sterile that animals raised in it will be less able, or willing, to display complex cognitive abilities, language abilities and consciousness.

Individuality and problems for testing

Throughout this book I have spoken sometimes of the characteristics of species and sometimes of the characteristics of individuals. At times I have been referring to those characteristics that are common to all, or at least most, members of a species. At other times I have been concerned to refer to the special characteristics of an individual and thus to recognise that, even within one species, individuals may differ. This is particularly evident when one looks at the individual as a whole, taking into account a large number of its characteristics. Thus, one can become aware of the individual as a separate self.

When an individual develops, it does so within a framework of experiences in a particular, although changing, environmental context. Within limits, it will be only in that environment that its sense of self might be fully expressed. If the self is not a self in isolation but one expressed within a particular social and physical context, that self may not be expressed in an alien environment. Thus, if we pluck an animal from the wild and bring it into the laboratory in order to test whether it has awareness, we may be defeating our purpose. The wild animal brought into the laboratory has to adapt to the presence of humans. It also has to adapt to the loss of other members of its own species. Such an individual would find its memory store to be of only limited use in directing its behaviour in the new environment. It would have lost a structure on which it could hang its sense of self.

These are massive changes, which must alter its sense of self and almost all of its cognitive patterns although, in time, adaptation and new learning would occur and a new sense of self may develop in the new context. Yet, often, wild-caught animals are tested along with animals raised in captivity with no exceptions made and nothing of their past history taken into account. Both captive-raised and wild-caught chimpanzees have been tested for self-recognition in tests using the mirror and red dye, outlined in chapter 2. The researchers have always stated this fact but it has not been considered in even the more comprehensive interpretations of the results.

Recognition of individual variation raises another problem so frequently encountered in this area of research. As I said previously, statements are often made about an entire species on the basis of results that have been collected from testing only a few individuals. Even worse, statements are often made about *all* animals on the basis of results collected from only a few animals and a few species. It is often said that animals cannot do something that humans can do. Humans are members of one species; the collective term 'animals' is used for the thousands of other animal species. Those thousands of species are not a unit that can sensibly be compared with humans.

Moreover, most animals differ one from another as much as do humans. We pretend that they are all alike. We also make comparisons between species of animals on the basis of results from very few representatives of each species. To give another example, a rather small number of capuchin monkeys have been tested in captivity on a task to be solved by using a tool (see chapter 3), the results are compared with those for the small number of chimpanzees that have been tested on a similar but not identical task, and the conclusion is reached that capuchins solve the problem by mindlessly trying every solution whereas chimpanzees contemplate it and use thought to reach a solution. Thus, in one sweep of the scientist's pen, all capuchins are condemned to a position behind the barrier of consciousness. This approach is not just unreasonable, it is unscientific. Scientists should take into account all of the factors that may influence their results. However, the problems created by not doing so are common in the field of animal cognition and awareness.

Perhaps this is about to change, as there are some scientists who have stressed the importance of individual differences. The primatologist Sarah Boysen of Ohio State University, USA, has said that the best description of the range of chimpanzee features and behaviour represented across chimpanzee populations in the wild and in numerous captive environments is remarkable variability.

I would say that the same is true of most other species. We all know this from the pets that live in close contact with us. No two dogs are the same, even when they are from the same litter, and the same can be said of cats, parrots and so on. It is the intimacy of knowledge of the pet owner that allows distinctions between individuals to be made. But animals do not change into being more uniform when they enter laboratories and become part of experiments. They may exhibit behaviours that they have in common, but they remain individuals. Scientists often forget this.

Learning from communication with other species

I have reasoned throughout this book that language (defined as the vocal communication used by humans; see chapter 1) is not an essential criterion for consciousness, although it certainly facilitates determining whether another individual is conscious. By means of language, it is possible to communicate what one is thinking about. Language is not essential to being conscious but it is a medium through which the mind can be expressed to another individual. The communication systems of other animals may, likewise, be used to express their minds but, so far, we have been unable to understand these systems well enough to see whether this is so. Instead, we have taught some animals to use our language.

Kanzi, a pygmy chimpanzee (or bonobo) at the Yerkes Regional Primate Research Center in Atlanta, USA, has been trained by Sue Savage-Rumbaugh to communicate by pointing at symbols on a board. He points at the symbols to communicate with humans but he can understand spoken English, and not in a trivial way of merely responding to commands; rather, it appears that he understands the syntax of the language. For example, if he is given the following instruction in pidgin English through ear phones 'Go get orange testing room' he will respond by going to get the orange, but he responds more rapidly and decisively when the syntactically correct command 'Go and get the orange from the testing room' is given instead.

I would suggest that other species that live in close contact with humans, such as our pet dogs, cats and birds, may understand aspects of language, provided that we have communicated with them in sensible ways that have meaning. Irene Pepperberg trained the parrot Alex by making sure that he overheard meaningful, simple verbal interactions between humans. For example, in front of Alex one person might ask whether the other has a key and the latter would say yes and hold up the key. Alex was not exposed to the meaningless 'Pretty boy', 'Polly want a drink?' phrases that we tend to say to birds. These phrases can be mimicked by parrots, and a number of other species of birds, but it is unlikely that they are understood by the birds because they have not been communicated to them in meaningful contexts.

I predict that many more species would understand aspects of our language if they were exposed to it in the same meaningful way as Alex and the signing apes have been, and with as much patience. The degree to which this might be possible will have to be determined and it is likely to vary amongst species and amongst individuals. I recognise that any research in this field is fraught with problems of training and interpretation, if it is to meet the strict criteria required to prove that an animal is producing or understanding language. Research on language in animals is of interest in its own right but, irrespective of this, we can test animals that have learnt to sign or to communicate with us in other ways to see how their minds work.

So far the focus of research with the animals that have been taught to communicate using English has been to find out whether or not they are actually using language, as we define it. It is not relevant to enter into the debate about their language abilities here. I would simply like to point out that there will be much more that we can learn from the signing apes, once the controversy about their language abilities is set aside and the researchers can get on with asking different questions. This is not to deny that there has been some attention paid to understanding the minds of the signing apes.

What do the signing apes tell us about their minds, quite apart from the issue of whether they use language or not? They signal desires, likes and dislikes and also memories of how they felt in the past. The lowland gorilla, called Koko, who has been taught to use sign language by Eugene Linden and Francine Patterson, at Stanford University, USA, has a working vocabulary of over 500 signs. She strings these signs together into statements of about three to six signs and she communicates about things in the present and past.

Koko had a companion kitten that died and the loss made her very sad. Later, when asked about it, she would express her sadness about the loss. When she saw a photograph of the kitten she again expressed her sorrow. She was able to communicate about the past and, therefore, think about an event not part of her immediate situation.

175

This is one of the criteria for consciousness, mentioned in chapter 1.

The signing apes also communicate about the future in terms of desires to go places or to be given things. Again, they display thoughts that are not part of their immediate situation. Perhaps the behavioural psychologist could find simple stimulus–response explanations for these acts of communication, but in my opinion the signing apes and speaking parrots open up the possibility of more exploration of their minds.

Asking an ape about its inner thoughts

Only from the animals that have learnt to communicate with us by signing or pointing to symbols can we expect to find out what they are thinking about. In the book *Kanzi: The Ape at the Brink of the Human Mind* Sue Savage-Rumbaugh describes an occasion in which she was riding in a car with Kanzi's sister Panbanisha. Noticing that Panbanisha looked as if she were lost in thought, Savage-Rumbaugh ventured to ask her what she was thinking about. The reply came after a few seconds of reflection and it was 'Kanzi'. Savage-Rumbaugh was surprised because Panbanisha rarely used the name Kanzi. Next Savage-Rumbaugh replied 'Oh, you are thinking about Kanzi, are you?', and Panbanisha vocalised excitedly in agreement.

Of course, this does not prove that Panbanisha was, in fact, thinking about Kanzi. Sometimes when we are asked what we are thinking about we respond with the first thing that comes into our minds. The same problem of reliability of information about spontaneous and private thoughts exists for animals and humans alike. I suspect that this is the reason why Savage-Rumbaugh has not often asked this question of the apes with whom she communicates. She also says that, when occasionally she has asked them what they are thinking, they usually ignore the question. However, it would be interesting to build up a larger repertoire of the answers to the question 'What are you thinking about?'.

One could compare the responses given to this question with those given to the question 'What are you dreaming about?' asked when the ape is awakened during rapid-eye-movement sleep, the phase of sleep in which dreaming occurs. If humans are awakened and asked what they are dreaming about, they can usually give an answer provided that they were in the rapid-eye-movement phase of sleep at the time. The waking makes the substance of the dream become conscious.

To my knowledge, no one has attempted to ask the sign-language-trained apes about their dreams, but it should be possible. What I am suggesting is that these responses should be compared with responses given when the ape is thinking and awake. If the two sets of answers are different for the most part, we may have an indication that they report genuine thinking because dreams and conscious thought, in humans at least, are rarely the same. Of course, if Panbanisha is particularly focussed on Kanzi, she might think about him in the day and also dream about him at night. In this case, similar sets of answers would not mean that apes do not have inner thoughts. However, a variety of answers and a difference between the sets of answers during the day versus the night would suggest that inner thought and dreams occur.

We need to ask important questions of the apes who have learned to communicate with us, and of parrots like Alex also. I agree with the following statement of Savage-Rumbaugh:

> To further our understanding of animal intelligence we must learn to ask better questions—questions that focus on unusual events, rather than mundane and readily controllable ones. If we were to start with the assumption that animals are conscious and capable of thought, reason, and complex communication, we would find it difficult to come up with evidence that would disprove this view. Instead, we start with the premise that they are incapable of such accomplishments

and find it difficult to disprove this view. (Savage-Rumbaugh and Lewin, 1994, pp. 263–264)

Brain waves and molecules of the mind

So far I have concentrated on measuring behaviour to understand the mind, and I have chosen to do this because the mind is expressed only in behaviour, whether that behaviour be language or something else. Thoughts require electrical activity in the brain and changes in the molecules inside the brain but these electrical and chemical events are not the mind itself. They are *correlates* of the expression of mind, but they do not embody the mind in its entirety, although many scientists researching these processes seem to believe that they do. It has recently become popular to use neurobiological approaches in the study of consciousness (i.e. to investigate physical and chemical aspects of nerve cell functioning in the brain) and, as so often happens, the scientists taking this approach forget that they are looking only at correlates of consciousness. Before long they begin to believe that the particular wave forms or chemical events that they are measuring *are* consciousness, and that way of reducing consciousness takes us away from the behaviour of the whole animal in the real world. The approach is called reductionism. It is an approach that runs the risk of forgetting that consciousness exists at higher levels of organisation and can only be expressed by the behaving, whole animal.

I was somewhat dismayed to find that a conference entitled 'Toward a Science of Consciousness' held in Tucson, USA, in 1996 was almost entirely devoted to nerve cell connections, molecular events, quantum mechanics of nerve cell function, computer modelling and some philosophy of the mind. Human perception was included, and some research that had used animals to record various chemical and physical aspects of brain function was reported. The latter had tested monkeys squatting in front of video monitors to measure eye movements. There was no paper that vaguely approached the theme of consciousness in animals. While I do not wish to detract from the challenging papers presented by eminent scientists in their particular fields, I do wish to express my surprise at the narrow focus of a conference aimed 'toward a science of consciousness'. It is as if it is acceptable to use animals to study the nuts and bolts of cognitive processes, whereas it is unfashionable, in these realms at least, to consider the expression of thinking in animals.

I am not about to criticise the direction of any research that deals with the baffling question of consciousness but I do believe that the study of consciousness should be broad enough and be approached open-mindedly enough to expand our visions, rather than working within the close confines of the constructs already in place. In my opinion, there is much to be gained from exchange of ideas and methods of study by considering the consciousness of humans and animals together. I would go a step further to say that there is much to be gained by comparing different species of animals, taking into account their differences and using them to illuminate the problems. Focussing on the mental processes of primates and ignoring those of birds has led many scientists to distorted views of the brain structures that might be involved in awareness and consciousness (chapter 4) and has provided a narrow view of evolution. Much can be achieved by comparing species even though we might do so merely to shed light on our own species. I would hope that there will be increased efforts to understand other species as well as our own.

Easy and hard problems of consciousness

Topics related to consciousness that have been covered in this book include clever or intelligent discrimination and categorisation of objects and events; cognitive integration of information and where it might occur in different parts of the brain; ways of responding that might reveal internal mental states such as self-awareness and awareness of the mind state of others, and communication about events of the past and of the conceived future. The mathematician and philosopher David Chalmers believes that these are the 'easy' problems of consciousness because they can be tackled by standard methods of science. Perhaps that is true if one confines the discussion to consciousness in humans but, as we have seen, these become 'hard' problems when we apply them to animals. The methods that we need to apply are neither standard nor easy.

Chalmers says that one of the hard problems of consciousness is the subjective *experience* of being conscious. We experience being able to see—for example, redness has a quality that we 'feel'—we experience emotions as an internal feeling; and we experience our train of thought. Philosophers call these qualitative feelings 'qualia'. As Daniel Dennett points out in his book *Consciousness Explained*, the conscious mind not only witnesses colours, smells and so on, but also *appreciates* them.

The qualia must arise from the workings of the brain, the electrical signals and the molecular changes and so on, but we do not know how. The problem with qualia is that they are completely private experiences and we do not know how *experience* of thinking comes about in humans, let alone in animals. Nor do we know how we might go about investigating the actual conscious experience. In chapter 2, I asked whether the young chick who is making distress calls actually has the experience of *feeling* distressed. We do not know of a way to access that feeling itself, if it exists, but we might assume that it exists in some form or another if we can demonstrate that the chick shows other characteristics of awareness. Those who seek to understand the subjective experience of consciousness will not be satisfied with the kind of research being carried out on awareness or consciousness in animals, but those who seek to learn more about animals will be excited by answers to the questions that Chalmers calls 'easy'.

CHAPTER SEVEN

THINKING, FEELING AND ANIMAL RIGHTS

We patronise them for their incompleteness, for their tragic fate of having taken form so far below ourselves. And therein we err, and greatly err. For the animal shall not be measured by man. In a world older and more complete than ours they move finished and complete, gifted with extensions of the senses we have lost or never attained, living by voices we shall never hear. They are not brethren; they are not underlings; they are other nations, caught with ourselves in the net of life and time, fellow prisoners of the splendour and travail of the earth. (Beston, 1971, pp. 19–20)

Throughout this book I have drawn attention to the ways in which our attitudes to animals have shaped our views and expectations of their cognitive abilities and awareness. The scientific study of animals is itself far from free of these attitudes. In this final chapter, it is appropriate to deal with the ways in which attitudes to the mental abilities of animals influence how we treat them and how we view them in the natural environment.

We have seen that species adapt to their particular environments. Most are, indeed, uniquely specialised to suit their own natural environments. But are most of them really so different from us? Intelligence and consciousness may have evolved many times over but the outcome might be functionally much the same.

Also, will we ever hear their voices? Beston (quoted

above) speaks as a naturalist in awe of the animals and natural environment that surrounded his cottage on the great beach of Cape Cod in Massachusetts, USA. His is a sentiment that many share and I must admit to times when I have been moved to think likewise. But this mystical concept of animals can be no more than a source of inspiration to seek more knowledge about them. As a scientist who studies the behaviour of animals, I do believe that we are coming closer to hearing the voices of other species and that their communication may not be beyond access by us. To reach it we will need a different perspective and a desire to understand, in the true sense of the word, not merely to exploit them for the purposes of humans. I am afraid that most funded research is for the latter category and relatively little support is given to understanding those other species that are 'caught with ourselves in the net of life and time'.

The issue is two-pronged. Unless we study them now, many animals will be no longer with us 'in the net of life and time', as they will be extinct. To recognise the need to study their behaviour, not merely for exploitation, will mean to change attitudes, to dismantle the divide that we have constructed between them and us.

Animals as individuals and identities lost

In chapter 6, the need to take individual differences between animals into account in research was discussed. Of course, it would be incorrect to say that individual differences apply equally to all species of animals from unicellular organisms to apes. No two animals are exactly the same, but individuality in brain function and behaviour must have become increasingly elaborate during evolution.

Physical and mental uniqueness of individuals might be a precursor to self-awareness because the self must be distinguishable from others. Social behaviour also relies on individuals being different. Each individual must be recognised by its appearance and behaviour.

If animals were merely machines, all members of the same species might be alike. Few of us consider this to be so for species with which we are familiar as pets or as working animals. Nevertheless, these experiences with animals on an individual level seem to do little to change our attitudes to animals in a general sense. We still tend to see species that are less familiar to us as unitary entities and to ignore individual differences, instead concentrating on the characteristics shared by all members of the species. We tend to treat species with which we are less familiar as invariant units. We give our pets names but think of wild species collectively 'as the kangaroos', 'the horses', and so on. The same is true in scientific research. Almost all research on animals involves testing animals in groups and average (mean) scores are calculated to represent the group, or even the species. As Lynda Birke of Warwick University, UK, has pointed out, scientists often think that they are working on standardised groups of animals (such as rats or rabbits) merely because they have not bothered to get to know the individuals well enough. Rats are as individually different from each other as are dogs and cats. Our attitudes are often a matter of convenience for research. This approach of studying animals as species has been useful, up to a point, in disciplines such as ethology and ecology, but even in these fields some researchers are starting to take individual differences into account. As I have discussed in previous chapters, taking into account individual differences is extremely important when one is studying complex behaviour, in particular behaviours that reflect consciousness. We need to live in close contact and communication with animals if we are going to be able to detect their subtle behaviours and if we are going to understand them in any way.

Societies of the past that lived in close contact with animals, either as hunters or as farmers on small farms with few animals, were acutely aware of the individuality of animals. This began to change with the advent of larger herds and flocks. By mediaeval times animals were seen as types or symbols and they could bear guilt in public ceremonies of punishment. The perceived characteristics of each type of animal were associated with the nature of the crime. In Europe, for example, dogs were hung on either side of a person who had committed a crime of immense infamy. Animals themselves were tried and punished if they inflicted harm on humans. In these times animals and humans lived in close association. In fact animals were, in some senses, seen as equivalent to humans, but still they were considered to be outsiders. They were seen to reflect humanity but to be outside it in a way that set a boundary between animals and humans. Although individual animals could be punished for human-type crimes by public executions the same as those used for humans, they were not seen as individuals but as species-specific types. 'Renard the fox', for example, epitomised a host of unacceptable traits in humans.

While those working on smaller farms may have, over the centuries, maintained close relationships with individual animals, as human society has increased in size and farming has become an entrepreneurial practice with ever-increasing sizes of flocks and herds, it has become impossible for farmers to know animals as individuals. In the industrial farming of today the identities of individual animals are completely lost. Animals in intensive farms are seen as bodies, to be fattened or to lay eggs. Knowledge of their behaviour is of concern only to prevent them from inflicting injury on each other or themselves, to stop feather pecking, tail and ear biting, and so on. Their higher cognitive abilities are ignored and definitely unwanted. I ended my previous book, *The Development of Brain and Behaviour in the Chicken*, with the statement that the domestic chicken is the avian species most exploited and least respected. Despite their domestication, chickens have retained complex cognitive abilities. They are not the same as feral or wild chickens, but the view of domestic chickens as stupid has more to do with how *we* think of chickens than with the abilities of the chickens themselves. The examples of communication behaviour and decision making in domestic chicks that we discussed in chapters 2 and 3 show that they are anything but stupid.

According to Peter Singer, a philosopher at Monash University, Australia, and author of *Animal Liberation*, the main issue underlying the construction of a gulf between animals and humans is to justify the eating of animals. Industrial farming relies on this gulf between 'them' and 'us'. So too does the new move into producing 'designer animals', ones genetically engineered to grow faster or produce a certain sort of meat or any other product that the market demands. Designer animals will still have minds, maybe even consciousness, but they will not be treated as such.

Do domestic animals have lesser minds?

An ultimate aim of breeding programs for domestic animals is to obtain animals that have minds so blunted that they will passively accept overcrowded housing conditions and having virtually nothing to do but eat—and then to eat standard and boring food delivered automatically. There is no evidence that domestic chickens, or other domestic breeds, have been so cognitively blunted that they need or want no more behavioural stimulation than they receive in battery farms. In fact, if domestic breeds are reintroduced to more natural conditions and bred there, they adapt rapidly to the better conditions. It is possible to change some aspects of behaviour by selective breeding but only within limits. Domestic breeds may be more docile, or less fearful and more accepting of the presence of humans, but these behaviours reflect temperament and motivation, not cognitive abilities.

In earlier chapters, I have pointed out the importance of environment on the development of brain and behaviour. No animal raised in captivity of any form, whether it be in intensive farming, a laboratory or a zoo, can adapt immediately to feral or wild conditions. In most cases, a prolonged period of rehabilitation training is required and in some cases adaptation to the new conditions may be impossible. This does not mean that the breed itself has shifted away from a need for more natural or more stimulating conditions. Domestic hens taken from battery cages may take some time to adapt to more freedom, but if their chicks are raised in more natural conditions they show surprising similarity to wild chickens. The cognitive capacity of the breed and its ability to perform complex behaviour appears to remain intact, despite generations of breeding under the control of humans.

Consciousness and animal welfare

Whether we assume that animals do or do not have consciousness determines how we treat them. Hence, cognition and consciousness in animals is unquestionably an issue of great importance to the welfare of animals, not only in research but also in other areas in which humans exploit animals.

It has been important to the entire Animal Welfare movement that scientists are beginning to accept that at least some species of animals (most guidelines apply only to vertebrate animals) can experience pain after the individual has reached a particular stage of development. Furthermore, many scientists now recognise that the pain felt may be somatic or psychological and that it may be specific to an individual, based on that individual's past experience and particular needs. Past experience with particular people can also be remembered and alter the amount of stress suffered by animals in subsequent experiments or procedures. The presence of a preferred human relieves stress, whereas one that is disliked exacerbates it. Memories of past events and associations become part of the present situation and compound the animal's feelings. The sensation of pain is not absolute but subjective and dependent on many different factors.

The sensation of pain is not directly related to awareness of self or of others, but awareness and consciousness might alter the kind of pain that is suffered and the subjective experience of pain. Although it is unquestionable that animals with consciousness will experience pain, failure to find evidence of consciousness in a species should not be used as a reason to conclude that the species does not feel pain. It must be remembered how difficult it is to design experiments that, in any way, measure consciousness. Added to this, there are the likely differences in consciousness between species, as well as between individuals. Since, as I have reasoned in this book, consciousness in its various manifestations may have evolved many times over, and thus species may have different intelligences and different forms of consciousness, it follows that the way in which animals experience pain may also vary from one species to the next. Although we have not yet found a way of establishing whether this is a fact, this way of viewing the experience of pain by animals provides a useful basis for animal welfare.

Marian Dawkins of Oxford University, UK, has said that decisions about whether an animal can feel pain do not have to be based on absolutes. One does not have to choose between the animal being, on the one hand, an automaton without consciousness and, on the other hand, having all of the elements of consciousness (as in humans). That is, the choice is not between an animal that is completely without an ability to feel pain and an animal that has the total sensation of pain, as we know it. The problem with this line of reasoning is that it places animal species on a hierarchical scale with humans at the top. Some animals are seen as having more elements of consciousness than have other animals, but rather than being a matter of more versus less it may be one of different kinds of consciousness in different animals. Thus the sensation of pain may be an issue not of more versus less pain but of different pain in different animals. Some support for this conceptualisation comes from the fact that humans experience different kinds of pain. For example, we can experience dull continuous pain versus rapid sharp pain, and these sensations are detected by different receptors and nerve endings in the skin and transmitted to the brain via different neural pathways. There are many other kinds of pain, some of which may be different degrees of the same kind of pain and others that are different sensations that we still refer to as pain. The one that might concern us most here is psychological pain. For example, we refer to the pain of loss, felt after a close friend dies. Koko, the gorilla who communicates using American Sign Language, expressed the same feeling of loss after her kitten died (chapter 6). Dogs have been known to pine away and die after the death of a human companion. There are many such examples, although the pain of loss has not been studied scientifically.

We also experience pain or suffering by seeing others suffer because we empathise with them. If animals can attribute mental states to others, as indeed we have strong indications that at least some species can (discussed in chapter 2), then we have to consider that an animal may suffer by seeing the suffering of others.

The gorilla Koko has demonstrated clearly that she can assess the suffering of others and feel sadness on their behalf. Koko has signed 'Sad?' when one of her carers expressed sadness. When Koko was shown a photograph of another gorilla struggling to get away from being bathed, she signed 'Me cry there', which suggests recognition of the picture and self-related identification rather than empathy. Empathy was shown in other situations: when her companion gorilla, Michael, was crying because he wanted to be let out of his room, Koko signed 'Feel sorry out'. There might be many animals thinking this in laboratory and farming environments as they watch or hear other animals being experimented on or being killed. None of the present guidelines for animal welfare take this into account.

There may be many levels of emotion and cognition that respond to seeing another member of one's species suffer. Species and individuals will vary on how this affects them, but we have no reason to believe that they are not affected. Legislation for animal welfare will, in my opinion, have to include guidelines for preventing suffering by empathy with the suffering of others.

At the present time legislation for appropriate caging conditions for animals used in research and agriculture takes into account the minimum requirements for the species. It is aimed to ensure that provision is made for the species to carry out its basic behaviours. Some say that a species must be able to express its 'instincts', innate behaviours. The debate has centred around, for example, whether battery hens should be given material in which they can dust-bathe or whether cattle in feedlots should be provided with shelter. These are such basic aspects of behavioural and physiological requirements that it can only be said that the debate is about providing minimally better housing conditions at the least financial cost. Once one begins to consider that the domestic animals in question have complex cognition and that they may require more stimulation than they receive in intensive farming conditions and in most laboratory housing, the debate about welfare moves on to an entirely new level.

Animals in confined caging or housed in conditions that provide them with little stimulation show stereotyped behaviours, meaning that they repeat the same behaviour over and over again. For example, pigs housed in isolation in crates or in overcrowded conditions with little to do will chew the bars or lick them in stereotyped ways. Animals in zoos frequently do likewise or they pace up and down along the walls of the cage. Humans in institutions, such as mental hospitals or gaols, also develop stereotyped behaviours. It seems that the stereoptypies provide some sort of physical stimulation, and perhaps some mental stimulation, that calms the stressed animals.

The conditions that are stressful vary with the species and the past experience of individuals, but being isolated is stressful to some and being overcrowded is stressful to others. Not being able to move around is clearly stressful to all animals. But what about stress caused by insufficient mental stimulation? We would not hesitate to accept that as a source of stress in humans (indeed, that is the basis of imprisonment and punishment), but few consider that it is the same for animals. It is now time that we took mental stimulation into our guidelines for animal welfare. It is already considered unacceptable to keep sheep in 'metabolism cages' (cages in which they stand on wire floors and that are so small that the sheep cannot turn around) for very long periods but this decision is based purely on their physical need to have exercise. The lack of mental stimulation that the sheep receives when confined in the cage may be just as stressful. Lack of stimulation is a recognised problem for pets, such as birds locked in cages with few things to play with or cats and dogs locked in the house while the owner goes out. There are pet therapy programs that seek to entertain animals in these situations. There are even video films available for cats and dogs to watch (bouncing balls and the like), but I know of no evidence that these actually provide the required stimulation. We know that species from fish to birds and primates will attend to video images but we do not know what they might choose to watch for any period of time!

Moving the barrier: The Great Ape Project

Attitudes to the welfare of animals are various and, as we have seen, they are changing, and will continue to do so, in response to the new information on higher cognition in animals. Some people are in support of guidelines and legislation specifically to protect animals. Others feel that animals have rights that must be protected.

In 1993 the book *The Great Ape Project* edited by Paola Cavalieri and Peter Singer was published. It advocated that all of the great apes, including humans, should be put within one family, instead of the present categorisation that separates humans from the other great apes. This position is based in part on our genetic similarity to the other apes (our genes differ by up to only 2 per cent: see chapter 5) and in part on the new discoveries of the intelligence of chimpanzees, orang-utans and gorillas.

In fact, some of the signing apes have been tested on intelligence tests designed for humans. For example, Lyn Miles at the University of Tennessee, USA, has tested Chantek, a signing orang-utan, on the standard Bayley Scale for Infant Development, which is used to assess mental development of human children. The tests include building towers of cubes, folding paper into certain shapes and pointing to specific pictures. At twenty-four months old, Chantek's score was equivalent to that of a human child of 13.6 months. At five and a half years old his score was equivalent to that of a human child of almost two years old. Other indicators of mental development, including symbolic play, language comprehension and tool use, put the five-year-old Chantek at the level of a four-year-old human child. The gorilla Koko showed much the same relationship to human intelligence. On some types of questions Koko did better than human children of the same age: namely, in discrimination between 'same' and 'different' and in detection of flaws in a series of incomplete or distorted drawings. On other types of questions, such as those requiring precise coordination to fit pieces of puzzles together, she was not as good as human children. Some intelligence tests have as much to do with movement control of the fingers and hands as to do with problem solving using cognition, and the construction of an orang-utan's and of a gorilla's hand does not make it easy for them to put together pieces of puzzle designed for human hands.

I find these results impressive, particularly when one considers that the human standards with which Chantek and Koko were compared were average values calculated by assessing a large number of children raised in environments very different from theirs. Although Chantek and Koko were given much attention and training, their worlds were very different from those of the human children with

191

which they were compared. In some ways they received more attention and stimulation than most human children and in other ways they received less. In particular, they were not permitted free movement in the world at large. This could have curtailed at least some aspects of their mental development, although other aspects might have been enhanced. The point is that, not being humans and not being raised entirely like most humans, orang-utans and other apes cannot be compared meaningfully with humans by using the same test. I grant that these comparisons have served to impress people regarding their mental capabilities, but only because previously we have believed them to be so inferior to us. Moreover, Chantek and Koko are single representatives of their species being compared with the average human child. How typical are they of their respective species? Chantek might be a very intelligent orang-utan, whatever we might mean when we apply this concept to an individual, and Koko might be a very intelligent gorilla. Alternatively, they might not be especially intelligent compared with other members of their species.

Of course, it could be argued that Chantek, Koko and the other signing apes were, in fact, raised to some extent as middle class American children and therefore the intelligence tests used were appropriate for them. In a sense I agree with this, at least in comparison with other apes, but they were not raised exactly like a human child and they still exhibit behaviours and abilities that are typical of their species. The researchers working with Koko recognised this and cited the following example to illustrate the point:

Answers that seem perfectly plausible to a gorilla must sometimes be scored as errors on standardized tests. For instance, the Kuhlmann-Andersen Test has two questions with a distinct human bias. One question directs the child to 'Point to the two things that are good to eat.' The choices are a block, an apple, a shoe, a flower, and an icecream sundae. Koko picked the apple and the flower. Another question asked the child to point out where it would run to shelter from the rain. The choices were a hat, a spoon, a tree, and a house. Koko sensibly chose the tree. (Patterson and Linden, 1981, p. 124)

The questions are also culture dependent. I have a Balinese friend who eats certain types of flowers as delicacies and might respond similarly to the same question.

More studies such as this will enlighten our search to understand the minds of apes but we must remember that the intelligence tests used have been designed for humans, not orang-utans or chimpanzees or gorillas. Indeed, intelligence tests are problematic even within human populations. In fact, intelligence tests are actually designed for middle and upper class, Western children and they do not transfer as accurate measures of the intelligence of working class children or children of other cultures. It has been possible to design an intelligence test on which working class children perform with higher scores than middle class children. The test asks the children to solve different sorts of social problems and to have different background knowledge. Judging by this, there should be no difficulty in designing an intelligence test on which orang-utans perform better than humans, provided that we know enough about the behaviour of orang-utans in the first place. Here the reader might be reminded of the pigeons that performed better than humans on a task requiring them to match stimuli presented at various rotations, a problem based on the Eysenck IQ test (chapter 3).

Despite the very serious problems with the standard human intelligence tests, it is of interest to give them to the signing apes, provided that the answers are interpreted creatively in order to demonstrate how close the performances scores of human and apes can be. However, I must admit to an element of concern. When tested on these intelligence tests designed for humans, apes will always have lower scores than humans of the same age and I suspect that this is used to confirm our feeling of superiority. We have designed the tests so that they will do just that.

Of course, the Great Ape Project takes other cognitive abilities of the great apes into account, when it aims to shift the boundary that presently divides us from the other great apes and thus extend to these animals the rights that are presently limited to humans. In a pragmatic sense, I certainly support this move. On the one hand, there is an urgent need to protect the dwindling numbers of great apes that are still surviving in the wild from being poached to be eaten or sold as pets. On the other hand, they should not be exhibited in zoos or used for medical research. Huge numbers of them are presently used for these purposes, particularly in the United States.

We extend human rights to people who cannot talk, or have not yet learnt to talk, and to humans of all levels of IQ performance, and rightly so. Yet, as proponents of the Great Ape Project point out, it can be said that the great apes overlap with the range of human performance. Apes and humans differ in some characteristics and overlap in others. The overlaps are justification for not separating them from us.

The Great Ape Project has raised these and many other important issues. However, my support of the Great Ape Project is not given without some reservations. By shifting the boundary to allow apes into the same group as humans, we are still saying that 'some animals are more equal than others'. In this book I have emphasised the higher cognitive abilities of birds. The intelligence of some species of birds is, in many ways, equivalent to that of some species of primates, even the apes. Yet genetically they are far removed from us. What can we do about their rights? The same may, in future, be said of many other species. Are we to grant rights to only our closest genetic relatives? Are we to do so on the basis of intelligence or awareness, both of which are impossible to assess on any single criterion? Whatever attribute we choose, there will be the problem of placing a boundary dividing those species that we think have 'it' from those that do not.

The future of thinking in animals

The debates about the welfare and rights of animals will continue, reliant on new information about cognition and consciousness in animals. Attitudes will change and those changes will also be resisted by those who have most to gain by thinking of animals as little more than clockwork machinery.

For many years the study of consciousness was seen as an unacceptable topic for those who study the structure and function of the brain (neuroscientists) as well as for those who study the behaviour of animals (ethologists). It was tainted with the intangible, considered beyond parsimonious explanation. Consciousness does, indeed, defy explanation in the simplest possible terms. It demands conceptualisation at higher levels of complexity, even involving a touch of the mysterious. That is its challenge. The ethologist Patrick Bateson of Cambridge University, UK, has said that slavish obedience to the maxim of parsimony tends to 'sterilize imagination' and that some of the most interesting attributes of animal behaviour are thus almost certainly overlooked. I could not agree more. By ignoring the most interesting attributes of the behaviour of animals we not only diminish our own experiences but also diminish the existence of animals.

FURTHER READING

Chapter 1

Boakes, R. (1984) *From Darwin to Behaviourism: Psychology and the Minds of Animals.* Cambridge University Press, Cambridge.

Byrne, R.W. (1995) *The Thinking Ape: Evolutionary Origins of Intelligence.* Oxford University Press, Oxford.

Chalmers, D.J. (1995) The puzzle of conscious experience. *Scientific American*, Dec., 62–68.

Cheney, D.L. and Seyfarth, R.M. (1990) *How Monkeys See the World: Inside the Mind of Another Species.* University of Chicago Press, Chicago.

Darwin, C. (1965) *The Expression of the Emotions in Man and Animals.* First published in 1872. Republished by University of Chicago Press, Chicago.

Dawkins, M.S. (1993) *Through Our Eyes Only? The Search for Animal Consciousness.* W.H. Freeman, Oxford.

Gardner, B.T. and Gardner, R.A. (1969) Teaching language to a chimpanzee. *Science*, 165, 664–672.

Griffin, D.R. (1992) *Animals Minds.* University of Chicago Press, Chicago.

——(1984) *Animal Thinking.* Harvard University Press, Cambridge.

Kaplan, G. and Rogers, L.J. (1994) *Orang-utans in Borneo.* University of New England Press, Armidale.

McFarland, D. (1989) *Problems of Animal Behaviour.* Longman Scientific and Technical, Harlow.

Premack, D. (1986) *Gavagai! or the Future History of the Animal Language Controversy.* MIT Press, Cambridge.

Ristau, C.A. (1991) *Cognitive Ethology: The Minds of Other Animals.* Lawrence Erlbaum, Hillsdale, N.J.

Romanes, G.J. (1882) *Animal Intelligence.* Kegan Paul, London.

Savage-Rumbaugh, S. (1994) *Kanzi: The Ape at the Brink of the Human Mind.* John Wiley and Sons, New York.

Terrace, H.S. (1979) *Nim.* Alfred A. Knopf, New York.

Whiten, A. (ed.) (1991) *Natural Theories of the Mind: Evolution, Development and Simulation of Everyday Mind Reading.* Basil Blackwell, Oxford.

Chapter 2

Boesch, C. (1991) Teaching among wild chimpanzees. *Animal Behaviour*, 41, 530–532.

——(1993) Aspects of transmission of tool-use in wild chimpanzees. In K.R. Gibson and T. Ingold (eds) *Tools, Language and Cognition in Human Evolution.* Cambridge University Press, Cambridge, pp. 171–183.

Byrne, R.W. (1995) *The Thinking Age: Evolutionary Origins of Intelligence.* Oxford University Press, Oxford.

Cheney, D.L. and Seyfarth, R.M. (1990) Attending to behaviour versus attending to knowledge: examining monkeys' attribution of mental states. *Animal Behaviour*, 40, 742–753.

——(1990) *How Monkeys See the World: Inside the Mind of Another Species.* University of Chicago Press, Chicago.

Epstein, R., Lanca, R.P. and Skinner, B.F. (1981) 'Self-awareness' in the pigeon. *Science*, 212, 695–696.

Evans, C.S., Evans, L. and Marler, P. (1993) On the meaning of alarm calls: functional reference in an avian vocal system. *Animal Behaviour*, 46, 23–38.

Gallup, G.G. Jr (1970) Chimpanzees: Self-recognition. *Science*, 167, 86–87.

Gallup, G.G. Jr, Povinelli, D.J., Suarez, S.D., Anderson, J.R., Lethmate, J. and Menzel, E.W. (1995) Further reflections on self-recognition in primates. *Animal Behaviour*, 50, 1525–1532.

Gyger, M. and Marler, P. (1988) Food calling in the domestic fowl (*Gallus gallus*): The role of external referents and deception. *Animal Behaviour*, 36, 358–365.

Hauser, M. (1996) *The Evolution of Communication.* MIT Press, Cambridge.

Hauser, M., Kralik, J., Botto-Mahan, C., Garret, M. and Oser, J.

(1995) Self-recognition in primates: Phylogeny and the salience of species-typical features. *Proceedings of the National Academy of Sciences*, 92, 10811–10814.

Heyes, C.M. (1994) Reflections on self-recognition in primates. *Animal Behaviour*, 47, 909–919.

Manning, A. and Serpell, J. (1994) *Animals and Human Society: Changing Perspectives*. Routledge, London.

Marler, P. and Evans, C. (1996) Bird calls: just emotional displays or something more? *Ibis*, 138, 26–33.

Marten, K. and Psarakos, S. (1995) Using self-view television to distinguish between self-examination and social behaviour in the bottlenose dolphin (*Tursiops truncatus*). *Consciousness and Cognition*, 4, 205–224.

Munn, C.A. (1986) Birds that 'cry wolf'. *Nature*, 319, 143–145.

Parker, S.T., Mitchell, R.W. and Boccia, M.L. (1994) *Self-awareness in Animals and Humans: Developmental Perspectives*. Cambridge University Press, Cambridge.

Povinelli, D.J. (1989) Failure to find self-recognition in Asian elephants (*Elephas maximus*) in contrast to their use of mirror cues to discover hidden food. *Journal of Comparative Psychology*, 103, 122–131.

Povinelli, D.J., Nelson, K.E. and Boysen, S.T. (1990) Inferences about guessing and knowing by chimpanzees (*Pan troglodytes*). *Journal of Comparative Psychology*, 104, 203–210.

Povinelli, D.J., Parks, K.A. and Novak, M.A. (1991) Do rhesus monkeys (*Macaca mulatta*) attribute knowledge and ignorance to other? *Journal of Comparative Psychology*, 105, 318–325.

Povinelli, D.J. and Preuss, T.M. (1995) Theory of mind: evolutionary history of a cognitive specialisation. *Trends in Neurosciences*, 18, 418–424.

Premack, D. and Woodruff, G. (1978) Does the chimpanzee have a theory of mind? *Behavioural and Brain Sciences*, 4, 515–526.

Rogers, L.J. (1995) *The Development of Brain and Behaviour in the Chicken*. CAB International, Oxon.

Russon, A.E., Bard, K.A. and Parker, S.T. (eds) (1996) *Reaching into the Minds of the Great Apes*. Cambridge University Press, Cambridge.

Seyfarth, R.M. and Cheney, D.L. (1992) Meaning and minds in monkeys. *Scientific American*, December issue, 78–84.

Chapter 3

Barlow, H., Blakemore, C. and Weston-Smith, M. (eds) (1990) *Images and Understanding*. Cambridge University Press, Cambridge.

Boesch, C. and Boesch, H. (1990) Tool use and tool making in wild chimpanzees. *Folio Primatologia*, 54, 86–99.

Boysen, S.T. and Capaldi, E.J. (eds) (1993) *The Development of Numerical Competence: Animal and Human Models*. Lawrence Erlbaum, New Jersey.

Brewer, S.M. and McGrew, W.C. (1990) Chimpanzee use of a tool set to get honey. *Folio of Primatologica*, 54, 100–104.

Byrne, R.W. (1995) Primate cognition: Comparing problems and skills. *American Journal of Primatology*, 37, 127–141.

Calvin, W.H. (1994) The emergence of intelligence. *Scientific American*, October 1994, 79–85.

Chevalier-Skolnikoff, S. and Liska, J. (1993) Tool use of wild and captive elephants. *Animal Behaviour*, 46, 209–219.

Delius, J.D. (1985) Cognitive processes in pigeons. In G. D'Ydelvalle (ed.) *Cognition, Information Processing and Motivation*. Elsevier, Amsterdam, pp. 3–18.

——(1987) Sapient sauropsids and hollering hominids. In W. Koch (ed.) *Geneses of Language*, Brockmeyer Bochum.

Donald, M. (1991) *Origins of the Modern Mind: Three Stages in the Evolution of Culture and Cognition*. Harvard University Press, Cambridge.

Emmerton, J. and Delius, J.D. (1993) Beyond sensation: Visual cognition in pigeons. In H.P. Zeigler and B.J. Bischof (eds) *Vision, Brain and Behaviour in Birds*. MIT Press, Cambridge, pp. 377–390.

Goodall, J. van Lawick (1968) The behaviour of free living chimpanzees in the Gombe Stream Reserve Tanzania. *Animal Behaviour Monographs*, 1, 161–311.

Gould, J.L. and Gould, C.G. (1994) *The Animal Mind*. Scientific American Library, New York.

Humphrey, N.K. (1976) The social function of intellect. In P.P.G. Bateson and R.A. Hinde (eds) *Growing Points in Ethology*. Cambridge University Press, Cambridge, pp. 303–317.

——(1992) *A History of the Mind*. Chatto and Windus, London.

Jolly, A. (1966) Lemur social behaviour and primate intelligence. *Science*, 153, 501–506.

Jones, T.B. and Kamil, A.C. (1973) Tool-making and tool-using in the Northern blue jay. *Science*, 180, 1076–1078.

Kaplan, G. and Rogers, L.J. (1994) *Orang-utans in Borneo*. University of New England Press, Armidale.

Krebs, J.R. (1990) Food-storing birds: adaptive specialisation in brain and behaviours? *Philosophical Transactions of the Royal Society of London, B*, 329, 153–160.

Mackintosh, N. (1994) Intelligence in evolution. In J. Khalfa (ed.) *What is Intelligence?* Cambridge University Press, Cambridge.

Macphail, E.M. (1982) *Brain and Intelligence in Vertebrates*. Clarendon Press, Oxford.

Millikan, G.C. and Bowman, R.I. (1967) Observations on Galápagos tool-using finches in captivity. *The Living Bird*, 6, 23–41.

Pepperberg, I.M. (1990) Conceptual ability of some nonprimate species, with an emphasis on an African Grey parrot. In S.T. Parker and K.R. Gibson (eds) *'Language' and Intelligence in Monkeys and Apes*. Cambridge University Press, Cambridge.

Regolin, L. and Vallortigara, G. (1995) Perception of partly occluded objects by young chicks. *Perception and Psychophysics*, 57, 971–976.

Rogers, L.J. and Kaplan, G. (1994) A new form of tool use by orang-utans in Sabah, East Malaysia. *Folia Primatologica*, 63, 50–52.

Skutch, A.F. (1996) *The Minds of Birds*. Texas A and M University Press, College Station.

Visalberghi, E. (1990) Tool use in *Cebus*. *Folia Primatologica*, 54, 146–154.

Visalberghi, E. and Fragaszy, D. (1990) Food-washing behaviour in tufted capuchin monkeys, *Cebus appella*, and crabeating macaques, *Macaca fascicularis*. *Animal Behaviour*, 40, 829–836.

von Fersen, L. and Güntürkün, O. (1990) Visual memory lateralization in birds. *Neuropsychologia*, 28, 1–7.

Westergaard, G.C. and Suomi, S.J. (1995) The manufacture and use of bamboo tools by monkeys: Possible implications for the development of material culture among East Asian hominids. *Journal of Archaeological Science*, 22, 677–681.

Wrangham, R.W., McGrew, W.C., de Waal, F.B.M. and Heltne, P.G. (eds) (1994) *Chimpanzee Cultures*. Harvard University Press, Cambridge, Mass.

Chapter 4

Bayer, S.A. and Altman, J. (1991) *Neocortical Development*. Raven Press, New York.

Bonner, J.T. (1980) *The Evolution of Culture in Animals*. Princeton University Press, New Jersey.

Bradshaw, J.L. and Rogers, L.J. (1993) *The Evolution of Lateral Asymmetries, Language, Tool Use and Intellect*. Academic Press, San Diego.

Cowell, P.E., Waters, N.S. and Denenberg, V.H. (1997) The effects of early environment on the development of functional laterality in Morris maze performance. *Laterality*, in press.

Diamond, M.C. (1988) *Enriching Heredity: The Impact of the Environment on the Anatomy of the Brain*. Free Press, New York.

DeVoogd, T.J., Krebs, J.R., Healy, S.D. and Purvis, A. (1993) Relations between song repertoire size and the volume of brain nuclei related to song: comparative evolutionary analyses amongst oscine birds. *Proceedings of the Royal Society of London, B*, 254, 75–82.

Eccles, J.C. (1989) *Evolution of the Brain: Creation of the Self*. Routledge, London.

——(1992) Evolution of consciousness. *Proceedings of the National Academy of Science*, 89, 7320–7324.

Finlay, B.L. and Darlington, R.B. (1995) Linked regularities in the development and evolution of mammalian brains. *Science*, 268, 1578–1584.

Fitch, R.H., Brown, C.P., O'Connor, K. and Tallal, P. (1993) Functional lateralization for auditory temporal processing in male and female rats. *Behavioural Neuroscience*, 107, 844–850.

Hellige, J.B. (1993) *Hemispheric Asymmetry: What's Right and What's Left*. Harvard University Press, Cambridge, Mass.

Krebs, J.R., Clayton, N.S., Healy, S.D., Cristol, C.A., Patel, S.N. and Jolliffe, A.R. (1996) The ecology of the avian brain: food-storing memory and the hippocampus. *Ibis*, 138, 34–46.

Krubitzer, L. (1995) The organization of neucortex in mammals: are species differences really so different? *Trends in Neurosciences*, 18, 408–417.

Northcutt, R.G. and Kaas, J. (1995) The emergence and evolution of mammalian neocortex. *Trends in Neurosciences*, 18, 373–379.

Nottebohm, F. (1989) From bird songs to neurogenesis. *Scientific American*, February, 56–61.

Petersen, S.E., Fox, P.T., Posner, M.I., Mintun, M. and Raichle, M.E. (1988) Positron emission tomographic studies of the cortical anatomy of single-word processing. *Nature*, 331, 585–589.

Popper, K.R. and Eccles, J.C. (1977) *The Self and Its Brain.* Springer-Verlag, Berlin.

Rogers, L.J. (1995) *The Development of Brain and Behaviour in the Chicken.* Cab International, Oxon, UK.

Saito, N. and Maekawa, M. (1993) Birdsong: the interface with human language. *Brain and Development*, 15, 31–40.

Sawaguchi, T. (1992) The size of the neocortex in relation to ecology and social structure in monkeys and apes. *Folia Primatologica*, 58, 131–145.

Sperry, R. (1983) *Science and Morality.* Basil Blackwell, Oxford.

Vallortigara, G. and Andrew, R.J. (1991) Lateralization of response by chicks to a change in a model partner. *Animal Behaviour*, 41, 187–194.

Chapter 5

Bisazza, A., Cantalupo, C., Robins, A., Rogers, L.J. and Vallortigara, G. (1996) Right-pawedness in toads. *Nature*, 379, 408.

Calvin, W.H. (1991) *The Ascent of Mind: Ice Age Climates and the Evolution of Intelligence.* Bantam Books, New York.

——(1994) The emergence of intelligence. *Scientific American*, October, 79–85.

Changeux, J-P. and Chavaillon, J. (eds) (1995) *Origins of the Human Brain.* Clarendon Press, Oxford.

Corballis, M.C. (1991) *The Lopsided Ape.* Oxford University Press, Oxford.

Diamond, J. (1991) *The Rise and Fall of the Third Chimpanzee.* Vintage, London.

Elliot, D. and Roy, E.A. (eds) (1996) *Manual Asymmetries in Motor Performance*, CRC Press, Boca Raton.

Falk, D. (1992) *Braindance.* Henry Holt, New York.

——(1992) Evolution of the brain and cognition in hominids. *The Sixty-Second James Arthur Lecture on the Evolution of the*

Human Brain. The American Museum of Natural History, New York.

Fine, M.I., McElroy, D., Rafi, J., King, C.B., Loesser, K.E. and Newton, S. (1996) Lateralization of pectoral stridulation sound production in the channel catfish. *Physiology and Behaviour,* 60, 753–757.

Fullagar, R.L.K., Price, D.M. and Head, L.M. (1996) Early human occupation of northern Australia: archeology and thermoluminescence dating of Jinmium rock-shelter, Northern Territory. *Antiquity,* 70, 751–773.

Gibson, K.R. and Ingold, T. (1993) *Tools, Language and Cognition in Human Evolution.* Cambridge University Press, Cambridge.

Leakey, R. (1994) *The Origin of Humankind.* Phoenix, London.

MacNeilage, P.F., Studdert-Kennedy, M.G. and Lindblom, B. (1987) Primate handedness reconsidered. *Behavioural and Brain Sciences,* 10, 247–303.

Marchant, L.F. and McGrew, W.C. (1996) Laterality of limb function in wild chimpanzees at Gombe National Park: comprehensive study of spontaneous activities. *Journal of Human Evolution,* 30, 427–443.

Marchant, L.F., McGrew, W.C. and Eibl-Eibesfeldt, I. (1995) Is human handedness universal? Ethological analyses from three traditional cultures. *Ethology,* 101, 239–258.

McGrew, W.C. and Marchant, L.F. (1992) Chimpanzees, tools, and termites: Hand preference or handedness? *Current Anthropology,* 33, 114–119.

Noble, W. and Davidson, I. (1996) *Human Evolution, Language and Mind.* Cambridge University Press, Cambridge.

Rogers, L.J. (1980) Lateralization in the avian brain. *Bird Behaviour,* 2, 1–12.

Rogers, L.J. and Kaplan, G. (1996) Hand preferences and other lateral biases in rehabilitated orang-utans, *Pongo pygmaeus. Animal Behaviour,* 51, 13–25.

Sugiyama, Y., Fushimi, T., Sakura, O. and Matsuzawa, T. (1993) Hand preference and tool use in wild chimpanzees. *Primates,* 34, 151–159.

Swisher, C.C. III, Rink, W.J., Antón, S.C., Schwarcz, H.P., Curtis, G.H., Suprijo, A. and Widiasmoro (1996) Latest *Homo erectus* of Java: Potential contemporaneity with *Homo sapiens* in Southeast Asia. *Science,* 274, 1870–1874.

Toth, N. (1985) Archeological evidence for preferential right-handedness in the lower and middle Pleistocene, and its possible implications. *Journal of Human Evolution*, 14, 607–614.

Ward, J.P. and Hopkins, W.D. (1993) *Primate Laterality: Current Behavioural Evidence of Primate Asymmetries*. Springer-Verlag, New York.

Chapter 6

Barchas, P.R. and Barchas, J.D. (1977) Sociopharmacology. In J.D. Barchas, P.A. Berger, R.D. Ciaranello and G.R. Elliot (eds) *Psychopharmacology: From Theory to Practice*. Oxford University Press, Oxford, pp. 80–87.

Bennett, M.R. (1977) *The Idea of Consciousness*. Harwood Academic, New York.

Boysen, S.T. (1994) Individual differences in the cognitive abilities of chimpanzees. In R.W. Wrangham, W.C. McGrew, F.B.M. de Waal and P.G. Heltne (eds) *Chimpanzee Cultures*. Harvard University Press, Cambridge, pp. 335–350.

Chalmers, D.J. (1996) *The Conscious Mind*. Oxford University Press, Oxford.

Dennett, D.C. (1991) *Consciousness Explained*. Little, Brown and Company, Boston.

Hameroff, S.R., Kaszniak, A.W. and Scott, A.E. (eds) (1996) *Toward a Science of Consciousness: The First Tucson Discussions and Debates*. MIT Press, Cambridge.

Heyes, C.M. (1993) Anecdotes, training, trapping and triangulation: do animals attribute mental states? *Animal Behaviour*, 46, 177–188.

Horn, G. (1988) What can the bird brain tell us about thought without language? In L. Weiskrantz (ed.) *Thought Without Language*. Clarendon Press, Oxford, pp. 279–304.

Miles, H.L.W. (1990) The cognitive foundations for reference in a signing orangutan. In S.T. Parker and K.R. Gibson (eds) *'Language' and Intelligence in Monkeys and Apes*. Cambridge University Press, Cambridge, pp. 511–539.

Patterson, F. and Gordon W. (1993) The case for the personhood of gorillas. In P. Cavalieri and P. Singer (eds) *The Great Ape Project*. Fourth Estate, London, pp. 58–77.

Patterson, F. and Linden, E. (1981) *The Education of Koko*. Andre Deutsch, London.

Rose, S. and Rose, H. (1970) *Science and Society*. Penguin Books, Harmondsworth.

Savage-Rumbaugh, S. and Lewin, R. (1994) *Kanzi: The Ape at the Brink of the Human Mind*. John Wiley and Sons, New York.

Weiskrantz, L. (1995) The problem with animal consciousness in relation to neuropsychology. *Behavioural Brain Research*, 71, 171–175.

Chapter 7

Allen, C. and Bekoff, M. (1995) Cognitive ethology and the intentionality of animal behaviour. *Mind and Language*, 10, 313–328.

Bateson, P. (1991) Assessment of pain in animals. *Animal Behaviour*, 42, 827–839.

Bekoff, M. (1994) Cognitive ethology and the treatment of non-human animals: How matters of mind inform matters of welfare. *Animal Welfare*, 3, 75–96.

Beston, H. (1971) *The Outermost House: a Year of Life on the Great Beach of Cape Cod*. Ballantine Books, New York. Original edition: Holt, Rinehart and Winston, 1928.

Birke, L. (1994) *Feminism, Animals and Science*. Open University Press, Buckingham.

Cavalieri, P. and Singer, P. (eds) (1993) *The Great Ape Project*. Fourth Estate, London.

Dawkins, M.S. (1980) *Animal Suffering. The Science of Animal Welfare*. Chapman and Hall, London.

Huffman, M.A. and Wrangham, R.W. (1994) Diversity of medicinal plant use by chimpanzees in the wild. In R.W. Wrangham, W.C. McGrew, F.B.M. de Waal and P.G. Heltne (eds) *Chimpanzee Cultures*. Harvard University Press, Cambridge, Mass., pp. 129–148.

Ingold, T. (ed.) *What is an Animal?* Allen and Unwin, London.

Manning, A. and Serpell, J. (eds) (1994) *Animals and Human Society: Changing Perspectives*. Routledge, London.

Nicol, C.J. (1996) Farm animal cognition. *Animal Science*, 62, 375–391.

Povinelli, D.J., Nelson, K.E. and Boysen, S.T. (1992) Comprehension of role reversal in chimpanzees: evidence of empathy? *Animal Behaviour*, 43, 633–640.

Rogers, L.J. (1995) *The Development of Brain and Behaviour in the Chicken.* CAB International, Oxon, UK.

Rollins, B.E. (1981) *Animal Rights and Human Morality.* Prometheus Books, Buffalo.

——(1989) *The Unheeded Cry: Animal Consciousness, Animal Pain and Science.* Oxford University Press, Oxford.

——(1995) *Farm Animal Welfare. Social, Bioethical and Research Issues.* Iowa State University Press, Ames.

Singer, P. (1975) *Animal Liberation.* New York Review Press, New York.

Wood-Gush, D.G.M. (1983) *Elements of Ethology.* Chapman and Hall, London.

Wood-Gush, D.G.M., Dawkins, M. and Ewbank, R. (eds) (1981) *Self-awareness in Domesticated Animals.* The Universities Federation of Animal Welfare, Hertfordshire, UK.

INDEX

207